NMR 25

Basic Principles and Progress

NMR at Very High Field

Guest-Editor: J. B. Robert

With contributions by
E. W. Bastiaan, D. Canet, R. Freeman,
U. Haeberlen, C. MacLean, D. Marion,
J. B. Robert

With 44 Figures

Springer-Verlag
Berlin Heidelberg New York London
Paris Tokyo Hong Kong Barcelona

ISBN 3-540-52946-2 Springer-Verlag Berlin Heidelberg New York
ISBN 0-387-52946-2 Springer-Verlag New York Berlin Heidelberg

Typesetting Thomson Press (India) Ltd, New Delhi
Offsetprinting: Colordruck Dorfi GmbH, Berlin
Bookbinding: Lüderitz & Bauer, Berlin
2151/3020-543210 - Printed on acid-free paper

Foreword

In recent years several improvements have been made in the manufacturing of resistive, superconducting and hybrid magnets. Condensed matter physicists are nowadays doing experiments in steady magnetic fields of up to 30 Tesla. But the field homogenity $\{{}_{B}{}^{\mu B}\}$, required in a volume of the order of a few cm^3 is usually several orders of magnitude less severe than the one which is needed for high resolution NMR. Over the last 30 years, with each generation of new high resolution NMR spectrometers, from 100 MHz up to 600 MHz, taking advantage of the increase in sensitivity and resolution, new areas of research have been opened in chemistry, physical chemistry and biochemistry. The generation of the 20 Tesla superconducting magnets is coming. Thus one may seriously start to consider high resolution NMR at 1 GHz. The purpose of this volume is to examine some of the advantages which can be obtained at such high frequencies and some of the problems we shall be facing.

An important aspect of NMR at high field which is not presented in this volume concerns the design of the magnet. The building of a superconducting magnet, producing a field higher than 20 T, with a field homogeneity $\mu B/B \;\; 10^{-10}$, in a cm^3 volume still remains today in 1990 a major challenge.

Grenoble, France J. B. Robert

Guest-Editor

Professor J. B. Robert

Service National des Champs Intenses
CNRS, B. P. 166X
38042 Grenoble Cedex, France

Table of Contents

A Brief History of High Resolution NMR

Ray Freeman[1] **and J. B. Robert**[2]

[1] Department of Chemistry, Cambridge University, England
[2] Service National des Champs Intenses CNRS, Grenoble, France

Table of Contents

NMR Basic Principles and Progress, Vol. 25

1 Introduction

The idea that nuclear magnetic resonance might be used in chemistry seems to have been suggested by S.S. Dharmatti, a chemist who happened to be working with magnetic resonance physicists at Stanford University. It had just been shown that the resonance frequency of a given nuclear species in a given magnetic field was not constant (as had been fondly hoped) but depended on the chemical nature of the sample. It appeared that the motion of the extranuclear electron cloud slightly modified the intensity of the applied magnetic field B_0, as it were "shielding" the nucleus to an extent of a few parts per million

$$B = B_0(1 - \sigma)$$

Disappointed physicists named the effect the "chemical shift" and moved on to more promising lines of research. Dharmatti pointed out that if this was indeed a universal effect, then a simple organic compound would show several proton resonance frequencies. For example ethanol should show three proton lines. The experiment was tried [1] and a high resolution NMR spectrum obtained, three lines with intensities in the ratio 1:2:3. It is an interesting commentary on scientific specialization that one of the authors, Martin Packard (a physicist), remarked that he had never really believed the structural formulae written by chemists until he saw the ethanol spectrum.

The Stanford group realized that magnetic fields of unprecedented uniformity would be required to study the tiny chemical shifts associated with proton magnetic resonance. In an amazing tour de force, they constructed a permanent magnet operating at 30 MHz for protons with a very highly homogeneous field [2, 3]. Using a trick suggested by Felix Bloch, they improved the effective homogeneity by spinning the sample with a small air turbine so that the proton spins behaved as if the field had been averaged over a circular path. The resolution was 0.5 Hz. Even today, over thirty years on, it requires special efforts to achieve resolution significantly better than this. Re-examination of the ethanol spectrum at high resolution provided a good example of the multiplet structure due to spin–spin coupling and the first instance of the effects of chemical exchange (the hydroxyl proton).

A few far-sighted and courageous chemists (notably H.S. Gutowsky and R.E. Richards) set out to build their own magnets and high resolution spectrometers from scratch, 30 MHz permanent magnets being favoured at that time. Some of the Stanford researchers joined the new company Varian Associates and constructed a commercial high resolution spectrometer using an electromagnet system. Within a decade every self-respecting Chemistry Department would have a high resolution NMR spectrometer. Often the "converts" to this new technique were infrared spectroscopists, who soon realized that NMR would be a more powerful structural tool than infrared, because the parameters used to describe the NMR spectrum (chemical shifts, coupling constants and relaxation times) were more directly related to molecular properties (electron density, proximity and molecular motion). A high resolution NMR spectrum is much more than a "fingerprint" for a given molecular structure.

2 NMR Chirality

The various spectroscopies which reflect electromagnetic interactions are invariant under inversion or mirror plane symmetry. Thus the electronic, vibration, rotation, EPR or NMR spectra of the two enantiomeric forms of a molecule dissolved in an achiral solvent are identical. However the intermolecular interactions between molecules comprising a racemic mixture in an optically pure solvent are, perforce, diastereomeric. Thus the spectra of left and right forms of a molecule dissolved in a chiral solvent may exhibit differences which are expected to be slight. Such a difference may be observed in NMR spectroscopy [4]. This technique is extremely interesting and may be routinely applied to check the optical purity of a mixture or even to assign the absolute configuration of a chiral molecule [5]. Such a differentiation between the spectra of chiral molecules which may theoretically be observed in any kind of spectroscopy, has only been observed in NMR. Moreover, the spectral line splitting between the diastereomers is proportional to the applied magnetic field, thus studies related to the intermolecular chiral interactions become more interesting as the magnetic field increases.

3 Double Resonance

NMR involves coherent radiation and in this sense it differs from the rest of "classical" spectroscopy as practised during the 1960's. Experimentally it was relatively simple to perform double resonance—the simultaneous application of two radiofrequency fields. Spin decoupling was widely used to identify coupled groups of spins [6] while spin tickling provided a picture of the way the observed transitions were related to an energy level diagram [7]. The facility to saturate a given transition provided a direct method for monitoring chemical exchange [8] or cross-relaxation processes, the nuclear Overhauser effect [9]. Even forbidden multiple-quantum transitions were investigated on the early continuous-wave NMR spectrometers [10]. Analogous experiments in infrared and optical spectroscopy had to await the development of suitable tunable lasers.

4 Sensitivity

The Achilles heel was sensitivity. Energy levels of NMR systems are very close together and the Boltzmann population differences are small. Worse still, irradiation very quickly equalizes these populations (saturation) and the spin-lattice relaxation processes are slow. The obvious remedy was to increase the intensity of the magnetic field, since calculations suggested that sensitivity should increase as roughly the 3/2 power of the applied field.

This idea was all the more appealing because an increase in field intensity also increased the dispersion of chemical shifts, thus simplifying the spectra. A great deal of effort was expended in the late 1950s and early 1960s on the analysis of strongly-coupled spin systems (where chemical shift differences were comparable with the coupling constants). Nowadays, strong coupling effects are much less serious because the applied magnetic fields are an order of magnitude more intense. The problem of overlap of resonances with very similar shifts was attacked in 1969 by the introduction of the lanthanide shift reagents [11, 12]. Small quantities of these reagents, added to the sample, produced dramatic differential shifts of the various resonances through what is known as the pseudocontact interaction. It was almost as if the spectroscopist had the benefit of a much higher magnetic field.

Improving sensitivity was not simply a question of constructing a more powerful magnet. High uniformity and high stability were also required and the radiofrequency circuitry became more difficult at higher frequencies. The metallurgy and fabrication of the pole caps was a critical factor for high homogeneity. Field stability was greatly improved by the flux stabilizer which monitored changes in magnetic flux passing through a pick-up coil and fed back an amplified correction flux from a "buck-out" coil. This took care of all but the long-term drift of the magnetic field. This final problem was countered by using an NMR signal to control the absolute value of the magnetic field, the so-called "field/frequency lock". A crucial contribution here was made by Hans Primas [13] who showed that this lock signal could be derived from the internal reference material tetramethylsilane (TMS) added to all samples as a chemical shift reference point. Nowadays a similar effect is achieved with the deuterium signal derived from a deuterated solvent.

It was clear that NMR had enormous potential by virtue of the high information content, but that some interesting nuclei, for example carbon-13, would not provide an adequate signal to noise ratio. The technique of multiscan averaging provided a partial solution to this problem [14] and it was fortunate indeed that the digital electronics needed to implement this was just becoming available. But time averaging has a practical limit set essentially by the time available in an overnight data accumulation, and this was, in general, not enough to satisfy the demands of carbon-13 spectroscopy in natural abundance. That had to await a new break-through in the technology.

5 Fourier Transform Spectrometers

The magnetic resonance phenomenon was normally explained in terms of the phenomenological Bloch equations, which have both transient and steady-state solutions. The early NMR spectroscopists tended to feel more comfortable with the steady-state regime, where a continuous low-level radiofrequency field was scanned in frequency through the range of interest. This was reminiscent of the methods used in optical spectroscopy, and was straightforward to explain to the chemists who employed the technique. But it had been long understood that the transient free induction signal after a pulse and the continuous-wave, frequency-swept spectrum formed a Fourier transform pair [15]. There were indications from infrared interferometry that Fourier transformation of an interferogram provided better sensitivity than the conventional frequency-scanning method since the entire spectrum was excited all the time, the so-called Fellgett advantage [16]. An early patent disclosure [17] suggested that a similar advantage could be derived from pulse-Fourier transform NMR. One could imagine a continuous-wave NMR spectrometer where the excitation was achieved by a set of N modulation sidebands at fixed frequencies spread evenly across the region of interest, followed by N narrow-band phase-sensitive detectors. The theory predicted a sensitivity improvement as high as $N^{1/2}$. This line of thinking led Ernst and Anderson [18] to build the first pulse-Fourier transform high resolution spectrometer and they were able to demonstrate a sensitivity improvement approaching the theoretical formula

$$S_{FT}/S_{CW} = (F/\Delta)^{1/2}$$

where F is the spectral width and Δ is the width of a typical resonance line. This enhancement approaches two orders of magnitude for carbon-13 spectroscopy.

Initially there were severe practical impediments to Fourier transform NMR spectroscopy. The free induction signals recorded by Ernst and Anderson were transferred to IBM punched cards and hand-carried to a computer centre for Fourier transformation; turnaround time, 24 hours. Each attempt at phase correction extended the data processing by one day. It was therefore particularly fortunate that small laboratory minicomputers were just becoming available, allowing the transformation to be performed on-line on the spectrometer. Nevertheless a typical Fourier transformation required about 20 minutes, enough to try the patience of any self-respecting organic chemist. The answer was provided by the Cooley–Tukey fast Fourier transform algorithm [19] allowing the same process to be completed in about 20 seconds. Fourier transform spectrometers were now a practical proposition.

The Fourier transform revolution was soon complete. Not only was there the sensitivity advantage of roughly two orders of magnitude, but time-dependent NMR phenomena became much more readily accessible to investigation. This was because the act of monitoring the behaviour of the nuclei was now accomplished in a time of the order of 10–20 microseconds (the time for a 90° read pulse). The time resolution was therefore much higher than for the old-fashioned experiments involving slow frequency scans. Fourier transform spectroscopy provided easy access to spin–spin and spin–lattice relaxation times of all the components of multiline spectra [20] and greatly facilitated the study of chemically induced nuclear polarization (CIDNP) and chemical exchange by saturation transfer [8]. It was hard to conceive of any good reason for using the continuous-wave frequency-sweep method any more, given that minicomputers were not expensive components compared to the overall cost of a spectrometer. (The exception was the dynamic range problem associated with proton NMR in dilute aqueous solutions; this has given rise to an entire armoury of techniques known as solvent suppression.)

6 Low Field NMR

Although the advantage of recording NMR spectra at high fields was soon recognized, in order to simplify the spectral analysis and increase the sensitivity, Béné [21] emphasized the interest of recording the spectra at very low fields, and used an NMR spectrometer where the static B_0 was the earth's magnetic field [22], with a proton frequency of 1950 Hz. At such a low frequency the second order effects, which depend on the relative signs of the indirect spin spin couplings, become more important. Thus by recording the proton NMR spectra of triethylphosphite-sulphide ($S = P(OC_2H_5)_3$), the ^2J (POCH) and ^3J(POCCH) spin spin coupling constants were shown to be of opposite sign [21].

Another technique has been developed to take advantage of recording zero-field NMR spectra of powder samples, where only the dipole–dipole or quadrupolar interactions are operating [23, 24]. In zero magnetic field, there is no externally imposed direction in space, and a broad spectrum is observed where the geometrical information is lost. However this information may be retrieved in the following way. The sample is moved from a large polarizing field B_0 to a weak collinear field B_2 which is suddenly switched off to initiate evolution under the zero-field Hamiltonian. This yields high resolution NMR spectra from polycrystalline samples similar to those from a single crystal, and provides the desired geometrical information.

7 High Field NMR

The pioneers of high resolution NMR were quite happy if they could lay hands on a magnet of any kind at all. A 30 MHz spectrometer already seemed too much like heavy engineering for a chemistry laboratory, and the fledgling NMR spectroscopists were normally banished to some dreary basement room for fear that the magnet might otherwise crash through someone's ceiling. Radiofrequency techniques are relatively straightforward at 30 MHz and there was some reluctance to venture very much higher in frequency. However, it became evident that proton high resolution spectra were much simpler at high field because chemical shifts were more widely dispersed while the instrument linewidth remained roughly constant. Strong coupling effects became less important, reducing the need for tedious spectral analysis. While the sensitivity improvement alone might not have convinced the spectroscopist to spend more money for a higher field, the combination of spectral simplicity *and* sensitivity provided an irresistable incentive. An example is shown in Fig. 1, which is a simulation of the phosphorus spectrum of $(PCH_3)_5$ as a function of magnetic field intensity [25]. The low-field spectrum has been analysed using spin tickling [7] and the NMR parameters used to predict the high-field spectra.

The first decade of high resolution NMR spectroscopy saw a fairly rapid progression from 30 MHz to 100 MHz spectrometers as magnet technology

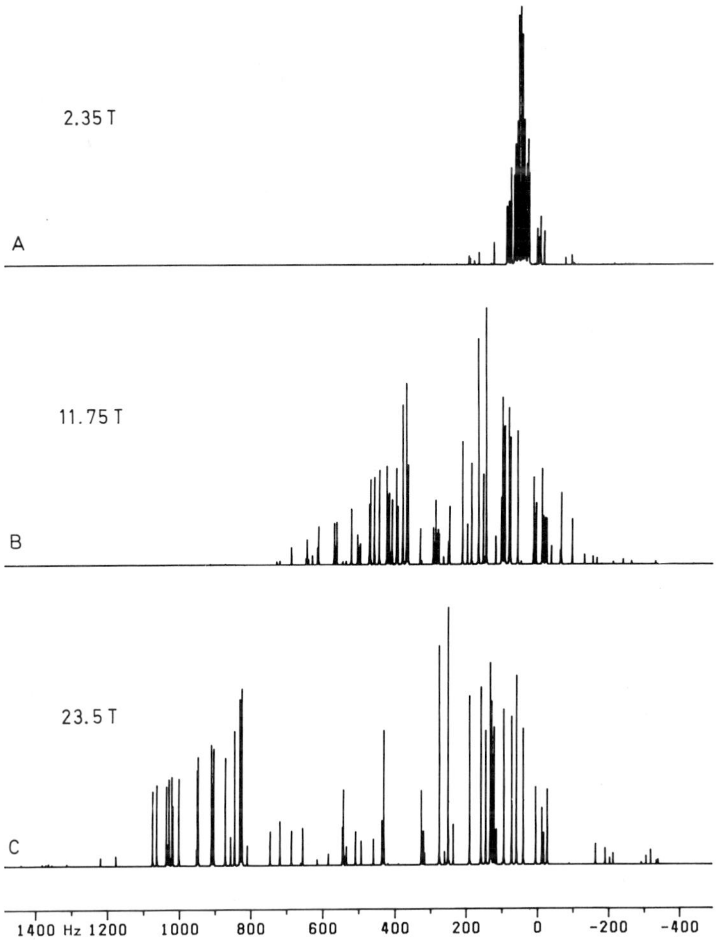

Fig. 1. Simulation of proton decoupled phosphorus spectra of pentamethylpentacyclophosphine at different field intensities top 2.35 T ($\nu\,^1H = 100$ MHz), centre 11.75 T ($\nu\,^1H = 500$ MHz), bottom 23.5 T ($\nu\,^1H = 1$ GHz)

improved. We can get a feeling for the progress achieved by plotting the proton resonance frequencies of commercial high resolution spectrometers against the year in which the first operating spectrometer was delivered to a customer (Fig. 2). The increase was approximately exponential (the scale in the figure is logarithmic, giving a straight line graph). The field intensity could not of course continue to increase if iron magnets were used, because of saturation effects, particularly in the pole caps. The practical limit was about 100 MHz. Permanent magnets reach saturation rather earlier and the highest fields (90 MHz) were only achieved by concentrating the flux from large diameter pole-pieces to smaller diameter pole-caps, an expensive solution.

Superconducting solenoids provide the answer [26]. The new type-II superconducting materials, for example niobium-tin, not only have zero electrical

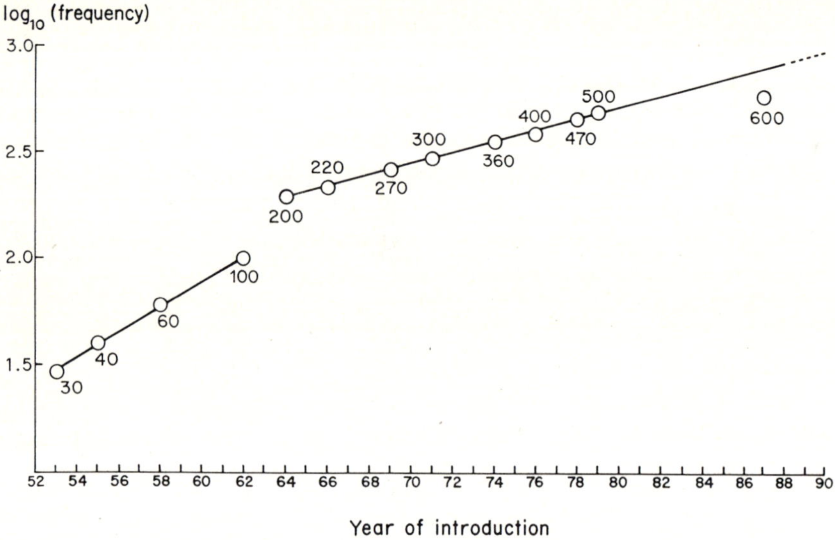

Fig. 2. Commercially available high resolution NMR spectrometers. A graph of the logarithm of the proton operating frequency against the year of the first installation in a customer's laboratory

resistance but also allow penetration of the magnetic field into the superconductor. They operate at high current and high magnetic field but have to be maintained below their critical temperature. In practice they operate at 4.2 K, the temperature of liquid helium, and are usually surrounded by a second chamber of liquid nitrogen to minimize helium loss. Closed-cycle refrigeration can also be used. The most gratifying technical improvement in recent years has been the careful redesign of the cryogenic dewar system to provide long helium hold times (about 3–4 months between helium refills). New superconducting materials are now available with much higher critical temperatures, well above the boiling point of liquid nitrogen. At the moment it appears that the magnetic and mechanical properties of these new superconductors are unsuitable for high-field NMR solenoids, but spectroscopists are nevertheless following these developments with great interest.

The graph shown in Fig. 2 suggests that field intensities of commercial superconducting solenoids for NMR also increased roughly exponentially with time but about twice as slowly as the graph for iron magnets. (Note that an *experimental* 600 MHz spectrometer was put into operation several years before the commercial 600 MHz spectrometer represented on this chart. Fields of this intensity require a new type of superconductor technology, and we may speculate that technical problems delayed the installation of the commercial machine). *If* the trend shown in Fig. 2 is still meaningful and *if* it continues, a 1000 MHz spectrometer should be introduced about the year 1990.

If one considers the possibility of developing high resolution NMR spectroscopy at higher frequencies, the value of 1 GHz for the proton resonance seems to be an upper limit at the present time. This corresponds to a magnetic field of about 23 T.

The highest steady magnetic field obtained with a resistive coil [27] (Bitter magnet) is 25 T, and with a superconducting coil [28] it is 20 T. With a hybrid system, which combines an external superconducting coil with an internal resistive coil [29], one may reach 31.5 T. By increasing the electric power available, several laboratories are presently considering the construction of systems delivering around 40 T in a steady state. However none of these fields are of NMR quality. To take advantage of the development of high magnetic fields in NMR for high resolution problems, one must keep the absolute resolution constant in frequency units. This means that a resolution of about 0.2 Hz must be maintained at 1 GHz($\Delta B/B \sim 2 \times 10^{-10}$). Thus a current stability close to 10^{-10} must be achieved. Up to a proton resonance frequency of 750 MHz, a superconducting magnet working in the permanent mode may be considered. At higher frequencies the permanent mode can no longer be used, owing to the number of superconducting joints required in the coil, and for safety reasons. Besides the high T_C superconducting materials, which are only available in small quantities, several other kinds of materials may be considered for the construction of a magnet of NMR quality at about 23 T:NbTi and Nb_3Sn up to 20 T, and V_3Ga, Nb_3Ge, Nb_3Al or Chevrel phases $PbMo_6S_8$ up to a field of 23 T.

Very high fields may turn out to be a mixed blessing. Already the proton resonance frequency is approaching the point where radiofrequency coils may have to give way to resonant cavities of the kind used for electron spin resonance. For studies on nuclei such as phosphorus-31, spin-lattice relaxation by the chemical shift anisotropy mechanism could become dominant at high fields, broadening the lines and thus negating the increased chemical shift dispersion. Finally, the assumption of completely isotropic molecular motion is not entirely justified at very high field. Even quite simple molecules experience a small amount of alignment, which may lead to very small additional splittings of the resonance due to dipole–dipole or quadrupolar interactions. This is a particularly serious problem if the internal field/frequency reference material is affected (deuterochloroform for example). In spite of these shortcomings, we can confidently expect that high resolution spectrometers will advance to higher magnetic fields in the next few years.

8 Ultrahigh Resolution

The efforts to reach higher and higher magnetic fields are justified by the resulting improvements in resolution and sensitivity. However, it must be pointed out that without changing the value of the applied magnetic field B_0, technical modifications have been developed to increase the resolution quite significantly [30]. Recently Allerhand and his coworkers considerably improved the resolution in liquid phase NMR spectroscopy, showing that the linewidth which may be reached on commercially available superconducting magnets is much better than previously believed [31]. If the temperature gradient within the sample may be lowered down to a few hundredths of a degree by increasing the flow of cooling air into the probe, and by decreasing the decoupling power [32] linewidths of less than 10 *milli* Hz may be obtained. If the T_2 relaxation time is not too short (chemical shift anisotropy), the ultrahigh resolution technique applied to high field NMR spectroscopy may become extremely interesting, allowing one to detect fine stereochemical changes or intermolecular interactions or even phenomena such as parity non-conservation in molecules [33] which has been predicted but never observed.

9 The Dipole–Dipole Interaction

In the early days of high resolution NMR the accepted wisdom was that solid samples had broad lines and were therefore unsuitable for study. Consequently (it was believed), high resolution studies should be restricted to liquid samples which had narrow lines because the dipole–dipole interaction was averaged to zero by the rapid molecular motion. The organic chemist was quite happy to work in the liquid phase. Nevertheless the thought remained that it would be most useful to be able to measure the dipole–dipole splitting since it is directly related to the inverse cube of the internuclear separation and therefore an ideal structural parameter.

It was soon demonstrated that this was one of those rare instances where we can have our cake and eat it. Saupe and Englert [34] showed that by dissolving a small molecule in a liquid crystal matrix (which becomes oriented in a magnetic field) it was possible to induce a small amount of alignment in the solute molecule even though it was tumbling rapidly and exhibited narrow lines. Scaled-down dipole–dipole splittings appeared in the spectrum, usually dominating the J-couplings and even chemical shift effects. The unknown parameter was the degree of alignment (which determined the scaling factor) but this could be calculated by *assuming* a value for one of the internuclear distances, provided the molecule was rigid [35]. Thus the method really measures the *ratios* of internuclear distances. The experimental accuracy is high since a 3% error in measuring a splitting produces only a 1% error in the internuclear distance. It is interesting to note that these spectra are not much improved by operation at high field (except with regard to sensitivity). Figure 3 shows the 200 MHz NMR spectrum of a sample of 3-phenylthiophene in a liquid crystal matrix (RCH) [36].

When we examine the other side of the coin, we realise that it was also an oversimplification to say that solid samples were unsuitable for high resolution work. Andrew et al. [37] showed that the dipole–dipole interaction in a solid could be removed by spinning the sample very rapidly at the "magic angle", 54° 44′. Thus the undesirable broadening effect from neighbouring spins could be suppressed, leaving finer details of the spectra to be studied, for example the anisotropy of the chemical shielding.

The technological problems of spinning a solid sample at very high speed were formidable. If the spinning rate was too low, a forest of spinning sidebands was observed and interesting features of the spectrum could be obscured. Occasionally at high spinning rates the rotors disintegrated explosively. These problems were

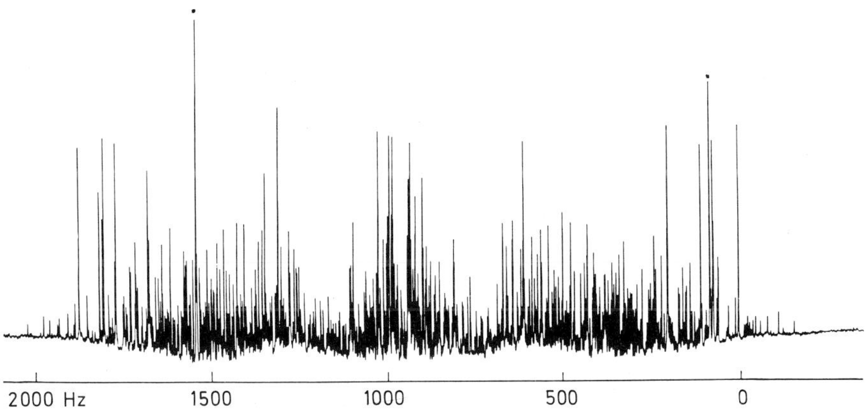

Fig. 3. 200 MHz proton NMR spectrum of 3-phenylthiophene dissolved in a liquid crystal matrix, showing a wealth of dipole–dipole splittings. The two *starred lines* are water transitions

gradually overcome by careful rotor design, the use of small rotors constructed of composite materials, and by using helium rather than air as the driver gas. Spinning rates as high as 15 kHz have been achieved in favourable cases.

Today, magic angle spinning experiments are often performed on nuclei of low natural abundance, such as carbon-13 [38], combined with proton–carbon cross polarization [39]. In this way the carbon-13 sensitivity approaches that of the protons (but still at 1% natural abundance). The method has been particularly successful for the study of polymers.

There is another way to achieve an effect analogous to spinning the sample; it averages the dipole-dipole interaction in spin space rather than in geometrical space. This entails the application of a sequence of very intense radiofrequency pulses calculated to "switch" the Hamiltonian between states we may call X, Y and Z, equivalent to spinning about the 111 axis (the magic angle). Clearly it is more practical to achieve very high spinning rates in spin space where there is no macroscopic motion. The pulse sequences that were used in practice [40–42] were often extremely complicated.

10 Broadband Decoupling

Carbon-13 spectroscopy only became practically viable when two new requirements were met—enhanced sensitivity and spectral simplification. Fourier transform spectrometers gave the improved sensitivity; broadband decoupling provided simplified spectra. The problem was to remove all proton–carbon fine structure over the entire range of proton chemical shifts, leaving only carbon-13 chemical shift effects. Coherent proton decoupling was only effective over a very narrow band of frequencies. The breakthrough was achieved by Ernst [43] with the concept of noise decoupling, in which a radiofrequency source underwent rapid phase inversions at pseudo-random intervals. No rigorous theory existed for the phenomenon of noise decoupling but it was thought to be analogous to the incoherent decoupling that occurs when the protons take part in rapid chemical exchange.

For many years noise decoupling was the accepted method and it worked extremely well. It was only when higher field spectrometers came into use (extending the range of proton chemical shifts) that limitations became evident. To achieve efficient decoupling the power level of the irradiation had to be increased and this gave rise to sample overheating, damaging some biochemical samples and giving rise to homogeneity problems due to convection in the sample. Various palliatives were tried, notably gated decoupling, square-wave phase modulation and chirp modulation. These were soon superseded by a new family of deterministic pulse sequences using the concept of composite pulses [44]. Starting with a composite spin inversion pulse R that was compensated for the effects of resonance offset, a magic cycle was assembled, $RR\bar{R}\bar{R}$, where $\bar{R}$ represents the composite pulse R with all radiofrequency phases inverted. Magic cycles were combined into supercycles and each stage of expansion improved the decoupling performance. A general theory of these broadband decoupling sequences was developed by Waugh [45]. One widely adopted scheme was WALTZ-16 which achieves the effect of conventional noise decoupling at far lower levels of radiofrequency power [32], [46]. Sequences for wider-band operation have also been developed. Because of the coherent nature of the modulation, these deterministic decoupling sequences give rise to weak "cycling sidebands" in the carbon-13 spectrum.

11 Two-Dimensional Spectroscopy

Commercial Fourier transform spectroscopy first became available in the early 1970s. This triggered a broad range of investigation of low-sensitivity nuclei, notably carbon-13 but including many nuclear species previously neglected [47]. The possibility of monitoring time-dependent phenomena tempted other spectroscopists into an orgy of measurements of spin–spin and spin–lattice relaxation times. High resolution NMR was undergoing one of its periodic revivals and there were many new experiments to be tried. In this climate, a startlingly new development went almost unnoticed. A Belgian physicist, J. Jeener, carried out the first experiments [48] on two-dimensional Fourier spectroscopy. The essential new ingredient was the introduction of an *evolution time* (t_1) prior to the usual acquisition time (t_2) when the free induction decay was collected. In a series of N consecutive experiments, the evolution time was incremented in small steps, so that each free induction signal differed from its predecessor because of the motion of the nuclear spins during t_1. The NMR data thus took the form of a two-dimensional array $S(t_1, t_2)$ and could be transformed with respect to t_1 and with respect to t_2. The result was a two-dimensional spectrum $S(F_1, F_2)$.

Several types of two-dimensional experiments are possible, but in retrospect the most fruitful appears to be the experiment first performed by Jeener [48],

$$90^\circ - t_1 - 90^\circ - \text{Acquisition } (t_2)$$

This is now known as homonuclear correlation spectroscopy (COSY). By allowing protons to precess at their chemical shift frequency ν_A during t_1, and by transferring a part of this coherence to a coupled site with chemical shift ν_X during acquisition, it creates a *cross-peak* in the two-dimensional spectrum centred on the coordinates (ν_A, ν_X). Observation of such a cross-peak establishes that the two proton sites are J-coupled. For complex organic and biochemical molecules, these COSY spectra provide invaluable information on proximity since very long-range proton couplings vanish within the linewidth and do not generate cross-peaks.

It is unfortunate indeed that Jeener never published his two-dimensional spectra, contenting himself with an oral presentation of the principles of two-dimensional spectroscopy at a Summer School [48]. For some years the idea lay virtually dormant until Ernst [49] alerted the NMR community to the broad scope and generality of two-dimensional spectroscopy, and developed several ingenious applications of the concept. The long-neglected field of multiple-quantum spectroscopy took on a completely new lease of life, since it was possible to monitor multiple-quantum precession frequencies by allowing the appropriate coherence to evolve during t_1, detecting the effects indirectly during t_2. These coherences of forbidden transition cannot induce an observable signal directly. One practical application of double-quantum spectroscopy was the "INADEQUATE" experiment [50] which detects and assigns directly-bonded pairs of carbon-13 nuclei, allowing the carbon framework of a molecule to be built up one link at a time.

Biochemists now make widespread use of the two-dimensional cross-relaxation experiment ("NOESY") because it provides valuable information on proximity [51]. Since the nuclear Overhauser effect falls off as the inverse sixth power of internuclear distance, the detection of NOESY cross-peaks in a two-dimensional spectrum indicates that the atoms concerned are within a few Ångstrom units in space. This places distance constraints on the many possible three-dimensional structures which the molecule might adopt. A typical two-dimensional spectrum may contain hundreds or even thousands of NOESY cross-peaks, leading to overdetermination of the structural problem, and minimizing the impact of several implicit assumptions of the method. When combined with molecular dynamics calculations, nuclear Overhauser measurements can be extremely powerful for determining three-dimensional structure in the liquid phase. Closely related experiments allow chemical exchange effects to be monitored [52].

The new concept of two-dimensional spectroscopy gave rise to a whole new field of one-dimensional experiments involving coherence transfer, modulated spin echoes, multiple-quantum coherence and spectral editing. This developed into a veritable cottage industry for the design of pulse sequences and their associated acronyms. In many instances the one-dimensional spectrum could be regarded as a projection of a hypothetical two-dimensional spectrum. The structural chemist was seldom satisfied with the conventional NMR spectrum alone, but went on to determine proton–carbon or proton–proton correlations, or to simplify the spectrum through editing techniques or multiple-quantum filtration. It became widely accepted that nuclear spins could be induced to perform all kinds of complex gymnastics which could be exploited to extract a great deal of useful additional information. Pulse sequencing became an art form.

12 In Vivo Spectroscopy

The new technique of magnetic resonance imaging [53] did not appear at first to impinge on high resolution NMR spectroscopy; as usual the physicists and chemists went their own separate ways. Then it was shown [54] that phosphorus metabolites could be monitored in vivo in a human patient by placing a small "surface coil" over the appropriate region (muscle tissue) to "focus" the radiofrequency excitation on a selected volume close to the coil. A high resolution phosphorus-31 spectrum was obtained and used as evidence that this particular patient was suffering from a rare disease known as McArdle's syndrome.

Thus for the first time it became feasible to monitor biochemistry within a human patient by a non-invasive method. Strictly, of course, the patient was "invaded" by an intense static field B_0, a weak radiofrequency field B_1 and (possibly) by switched field gradients dB_0/dx etc., but compared with existing clinical sampling techniques, and even compared with X-ray tomography, NMR was a very mild method indeed.

There was an immediate surge of interest in investigations of the kidney [55], the heart and the brain. The prime information would be the relative intensities of resonances of known metabolites such as phosphocreatine and adenosinetriphosphate, and their time development during (say) renal malfunction or cardiac ischemia. The inorganic phosphate signal was shown to shift slightly as a function of pH; this provided evidence for lactic acid production in muscle tissue after exercise.

At present several different spatial localization techniques have been explored, some employing applied static field gradients, others relying on the non-uniformity of the radiofrequency field. Arrays of several radiofrequency coils have been used to improve the localization of the "volume of interest".

Alternatively, similar biological information has been obtained by the technique of chemical shift imaging, where a separate image is displayed for each chemically distinct constituent. This is the bridge between in vivo spectroscopy and magnetic resonance imaging (which is not concerned with chemical shifts).

These high-resolution spectra are not limited to phosphorus-31 studies, although this is a particularly suitable nucleus because only a very limited number of phosphorus-containing compounds appear in the spectra. Carbon-13 suffers from intrinsically poor sensitivity but has the advantage that isotopically enriched materials can be introduced. High resolution proton NMR can be used, provided that the spectra are not too complicated.

In vivo high resolution spectroscopy is still in its infancy, and since human biochemistry is known to be extremely complex, we should not expect solutions to important medical questions immediately. But the prognosis for improving our understanding of human biochemistry and for monitoring drug therapy appears to be very exciting indeed. The practice of medicine may never be the same again.

13 Concluding Remarks

The most striking feature of this story is that high resolution NMR never reached a stable plateau of development; whenever it *seemed* to do so, someone discovered an entirely new aspect of the subject, triggering further explosive growth. When 60 MHz proton spectrometers came on the scene they made the existing 40 MHz machines seem very feeble indeed. Old inhibitions about solid-state NMR were swept away almost overnight by new ideas about cross-polarization and magic angle spinning. Fourier transformation achieved a major revolution in the way we thought about the NMR phenomenon, and incidentally opened the door to computer control of experiments. Two-dimensional spectra in the form of intensity contour maps became common currency among structural chemists. Small liquid samples in 5mm tubes gave way to human patients reclining on a table which slid smoothly into an enormous superconducting magnet.

The story is clearly not finished yet.

14 References

1. Arnold JT, Dharmatti SS, Packard ME (1951) J. Chem. Phys. 19: 507
2. Arnold JT (1956) Phys. Rev. 102: 136
3. Anderson WA (1956) Phys. Rev. 102: 151
4. Pirkle WH (1966) J. Amer. Chem. Soc. 88: 1837; Burligame TG, Pirkle WH (1966) J. Amer. Chem. Soc. 88: 4294
5. Pirkle WH, Beare SD (1967) J. Amer. Chem. Soc. 89: 5485
6. Bloom AL, Shoolery JN (1955) Phys. Rev. 97: 1261
7. Freeman R, Anderson WA (1962) J. Chem. Phys. 37: 2053
8. Forsén S, Hoffman RA (1963) J. Chem. Phys. 39: 2892
9. Kuhlmann KF, Grant DM (1968) J. Amer. Chem. Soc. 90: 7355
10. Anderson WA, Freeman R, Reilly CA (1963) J. Chem. Phys. 39: 1518
11. Hinkley CC (1969) J. Amer. Chem. Soc. 91: 5160
12. Sanders JKM, Williams DH (1971) J. Amer. Chem. Soc. 93: 641
13. Primas H, Fifth European Congress of Molecular Spectroscopy, Amsterdam 1961
14. Klein MP, Barton GW (1963) Rev. Sci. Instru. 34: 754
15. Lowe IJ, Norberg RE (1957) Phys. Rev. 107: 46
16. Fellgett PB (1951) Ph.D. Thesis, Cambridge University
17. U.S. Patent 3, 287, 629 issued to Varian RH, Filed August 1956
18. Ernst RR, Anderson WA (1966) Rev. Sci. Instr. 37: 93
19. Cooley JW, Tukey JW (1965) Math. Computation 19: 297
20. Vold RL, Waugh JS, Klein MP, Phelps DE (1968) J. Chem. Phys. 48: 3831

21. Duval E, Béné GJ (1967) Helv. Chem. Acta. 40: 501
22. Hochstrasser G (1961) Hel. Chem. Acta. 34: 189
23. Weitekamp DP, Bielecki A, Zax D, Zilm K, Pines A (1983) Phys. Rev. Lett. 50: 1807
24. Mullar JM, Thayer AM, Bielecki A, Zax DB, Pines A (1985) J. Chem. Phys. 83: 934
25. Albrand JP, Gagnaire D, Robert JB (1973) J. Amer. Chem. Soc. 95: 6498
26. Nelson FA, Weaver HE (1964) Science 146: 3641
27. Schneider-Muntau HJ (1984) J. de Physique Colloque C1, 45, C15
28. Turowski P, Shneider Th (1988) IEEE Transactions on Magnetics 24: 1063
29. Schneider-Muntau HJ, Vallier JC (1988) IEEE Transactions on Magnetics 24: 1067
30. Gray KW, Ozier I (1971) Phys. Rev. Lett. 26: 161; Kachurin AM, Fomichev VN (1981) Sov. Phys. JETP 53: 99
31. Allerhand A, Addleman RE, Osman D (1985) J. Amer. Chem. Soc. 107: 5809; Maple SR, Allerhand A (1987) J. Amer. Chem. Soc. 109: 6609; Allerhand A, Maple SR (1988) J. Magn. Res. 76: 375
32. Shaka AJ, Keeler J, Freeman R (1983) J. Magn. Res. 53: 313
33. Barra AL, Robert JB, Wiesenfeld L (1988) Europhys. Lett. 5: 217
34. Saupe A, Englert G (1963) Phys. Rev. Lett. 11: 462
35. Diehl P, Khetrapal CL (1969) NMR basic principles and progress vol 1, p 1
36. Chidichimo G, Liguori H, Longeri M, Veracini CA (1983) J. Magn. Reson. 51: 438
37. Andrew ER, Bradbury A, Eades RG (1958) Archs. Sci. Genève 11: 223
38. Schaefer J, Stejskal EO (1976) J. Amer. Chem. Soc. 98: 1031
39. Pines A, Gibby MG, Waugh JS (1972) Chem. Phys. Lett. 15: 373
40. Waugh JS, Huber LM, Haeberlen U, Phys. Rev. Lett. 20: 180
41. Mansfield P (1970) Phys. Lett. A 32: 485
42. Rhim WK, Elleman DD, Vaughan RW (1973) J. Chem. Phys. 59: 3740
43. Ernst RR (1966) J. Chem. Phys. 45: 3845
44. Levitt MH, Freeman R (1979) J. Magn. Reson. 33: 473
45. Waugh JS (1982) J. Magn. Reson. 50: 30
46. Shaka AJ, Keeler J, Frenkiel TA, Freeman R (1983) J. Magn. Reson. 52: 335
47. See for example Laszlo P (ed) (1983) NMR of newly accessible nuclei, Academic, New York
48. Jeener J (1971) Ampère International Summer School, Basko Polje, Yugoslavia
49. Aue WP, Bartholdi E, Ernst RR (1976) J. Chem. Phys. 64: 2229
50. Bax A, Freeman R, Frenkiel TA (1981) J. Amer. Chem. Soc. 103: 2102
51. Kumar A, Ernst RR, Wüthrich K (1980) Biochem. Biophys. Res. Comm. 95: 1
52. Jeener J, Meier BH, Bachmann P, Ernst RR (1979) J. Chem. Phys. 71: 4546
53. Lauterbur PC (1973) Nature 242: 190
54. Ross BD, Radda GK, Gadian DG, Rocker G, Esiri M, Falconer-Smith J (1981) New Eng. J. Medicine 304: 1338
55. Ross BD, Freeman D, Chan L (1986) Kidney International 29: 131

Molecular Orientation in High-Field High-Resolution NMR

E. W. Bastiaan and C. MacLean

Department of Physical Chemistry, Free University, De Boelelaan 1083, 1081 HV Amsterdam, The Netherlands

Table of Contents

At high magnetic fields a special feature of high-resolution NMR becomes important. It originates from the tendency of molecules, having anisotropic magnetic properties, to orient in the magnetic field of the spectrometer. As a result, anisotropic nuclear interactions are incompletely averaged and become manifest in the spectra of solutes in the mobile phases, which therefore may show additional fine structure. At currently available high fields, the effects of magnetic dipolar and electric quadrupolar spin couplings can be quite pronounced. NMR spectroscopists should therefore be aware of these alignment effects when interpreting their spectra. The study of magnetic field induced molecular orientation in NMR offers some interesting applications. Thus, information may be obtained about molecular susceptibility anisotropies and asymmetries, intermolecular interactions, aromaticity, solution conformations and quadrupole coupling constants. These possibilities in combination with the inherent

NMR Basic Principles and Progress, Vol. 25

simplicity of the technique make it an attractive tool for investigating properties and interactions of molecules in liquid solutions. It is expected that the use of supercritical fluid solvents and the future availability of higher magnetic fields in NMR, will greatly enlarge the range of possibilities of the method.

Information Note

A special feature of high-resolution NMR at high fields is the extra fine structure in the spectra as a consequence of the orientation of the molecules by the magnetic field of the spectrometer. The study of these effects, which arise from incomplete averaged magnetic dipolar and/or electric quadrupolar spin couplings, constitutes a tool for obtaining information about molecular properties and interactions in liquid solutions.

1 Introduction

Great efforts have always been made in NMR spectroscopy to achieve stronger magnetic fields. This serves two goals: an increased sensitivity, which allows shorter measuring times or smaller amounts of substance, and an improved resolution, by which spectral interpretation is facilitated. Nowadays, fields up to 14.10 T (^{1}H: 600 MHz) are commercially available and access to spectrometer fields of 11.75 T (^{1}H: 500 MHz) is possible already for many spectroscopists.

When dealing with liquid or gaseous samples at these high fields, the less well recognized phenomenon of molecular alignment comes within the limits of detection. The effect arises because the magnetic field of the spectrometer exerts orienting torques on all molecules of the sample that have an anisotropic magnetic susceptibility. For example, diamagnetic molecules tend to align in such a way that the induced magnetic moments are minimal (in absolute value). Since the effects of the orienting torques are counteracted by the Brownian motion, a very incomplete dynamic orientation remains. As a result of this partial alignment, residual couplings due to anisotropic nuclear spin interactions may show up in high-resolution NMR spectra at high fields. Thus, anisotropic shift contributions, anisotropic indirect spin–spin couplings, direct magnetic dipole–dipole interactions and electric quadrupole interactions can, in principle, influence the spectra [1–3]. In general then, intensities, line widths and separations of NMR lines will be modified, and even additional fine structure may be found. When not resolved, therefore, these effects complicate spectral interpretation and will be seen as a burden. Fortunately, for currently available high field spectrometers the orientation effects can be quite pronounced and when recognized can be used to advantage in e.g. peak assignments [9] and determinations of signs of indirect coupling constants [4–5]. Furthermore, interpretation of the residual anisotropic couplings gives extra information concerning molecular properties and interactions [1, 6].

In experimental studies of magnetic field induced partial orientation so far, the spectral effects of anisotropy of nuclear shielding [3, 7] and of the indirect coupling [7] have not been observed, since they are too small. Attention is therefore restricted to the direct magnetic dipole–dipole interaction between nuclear spins [1, 3] and to the interaction of a nuclear quadrupole with the local electric field gradient caused by its surroundings [1, 8]. In particular, deuterium quadrupolar interactions [9] and proton–proton and proton–carbon(13) dipolar interactions [4, 10, 11] have been investigated extensively. As an example, the 500 MHz proton NMR spectrum of the fluorene CH_2-group is given in Fig. 1. Due to the slightly preferential orientation imparted on the molecules by the static magnetic field of the spectrometer, this resonance shows as a doublet, owing to an incompletely averaged dipolar coupling between the CH_2-protons.

The magnitude of the observed anisotropic couplings in the spectrum (dipolar, quadrupolar) depends on a nuclear coupling constant, the molecular geometry and on the degree of alignment of the molecules (see Sect. 2). Consequently, NMR alignment experiments can serve as a tool to study molecular geometry [6, 12–14] and nuclear quadrupole coupling [6, 15]. The degree of alignment for an axially

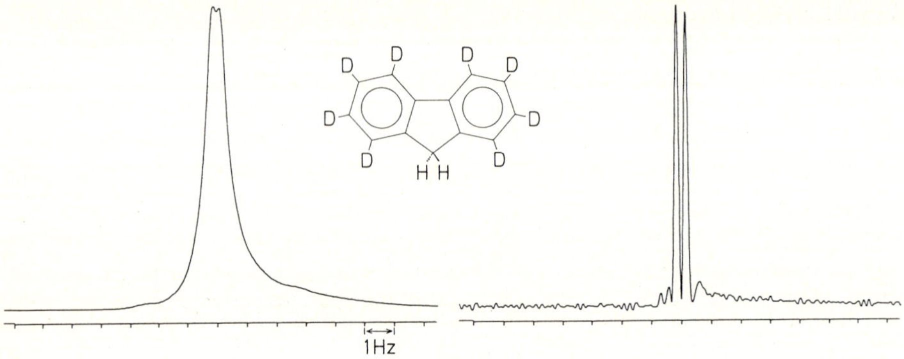

Fig. 1. The CH_2 resonance in the 500 MHz (11.75 T) proton NMR spectrum of 9H-fluorene-1,2,3,4,5,6,7,8-d_8 (1 mol % in acetone-d_6) is a doublet, due to the presence of a magnetic field induced dipolar coupling. The aromatic rings have been deuteriated to remove masking of the CH_2 doublet by small indirect couplings with the aromatic protons. The line splitting is found to be 0.30 Hz; the line width is about 0.60 Hz. The asymmetry on the right of the line is caused by a small quantity fluorene-d_9. Left: normal spectrum; right: resolution-enhanced spectrum

symmetric molecule is given by the average $\langle \frac{3}{2}\cos^2\theta - \frac{1}{2}\rangle$ [16, 17]. Here θ is the instantaneous angle between the principal molecular axis and some fixed external direction (for magnetic field induced orientation this will be the field direction); the brackets indicate averaging over the molecular tumbling. Completely oriented molecules have $\langle \frac{3}{2}\cos^2\theta - \frac{1}{2}\rangle = -1/2$ or $+1$; in the absence of average orientation $\langle \frac{3}{2}\cos^2\theta - \frac{1}{2}\rangle = 0$. Partial alignment induced by magnetic fields of 10–14 T typically leads to $\langle \frac{3}{2}\cos^2\theta - \frac{1}{2}\rangle_B \cong 10^{-6} - 10^{-5}$ for diamagnetic molecules and to $\langle \frac{3}{2}\cos^2\theta - \frac{1}{2}\rangle_B \cong 10^{-5} - 10^{-4}$ for paramagnetic molecules [1]; the subscript B indicates that the average is not an isotropic average, but is obtained in the presence of an aligning magnetic field **B**. As a result of this relatively small degree of alignment, the anisotropic effects in the spectrum range from less than 0.1 Hz to a few Hertz for diamagnetic compounds and from a few Hertz up to a hundred Hertz for paramagnetic substances. Whether these anisotropic couplings can be observed depends largely on the line widths in the spectrum. The degree of alignment $\langle \frac{3}{2}\cos^2\theta - \frac{1}{2}\rangle_B$ is proportional to the magnetic susceptibility anisotropy of the molecules and to the square of the magnetic field strength; it also depends on the temperature and it contains contributions from intermolecular interactions. Therefore, NMR magnetic field alignment studies in principle provide a means to obtain magnetic susceptibility anisotropies [1] and to investigate molecular correlation effects [18–20].

It will be clear from the foregoing, that when very high field NMR spectrometers are available (e.g. 23.5 T, ^{1}H: 1 GHz), the effects of magnetic field induced molecular orientation are unavoidable. It is the aim of this chapter, therefore, to make NMR spectroscopists aware of this phenomenon. First, in Sect. 2, it is outlined how the working equations for the residual dipolar and quadrupolar couplings are obtained. In Sect. 3, the accurate determination of the anisotropic coupling parameters is discussed. Applications are presented in Sect. 4: practical consequences of the alignment effects are pointed out and studies of magnetic susceptibilities, angular correlation effects, aromaticity, solution conformations and quadrupole coupling constants are reported. Section 5 gives some conclusions.

2 Theory

2.1 Incomplete Averaging of Anisotropic Spin Interactions

NMR spectra are usually interpreted in terms of a spin Hamiltonian that represents all nuclear spin interactions within an isolated molecule. These spin interactions are described by second rank tensors and depend on the molecule's orientation with respect to some imposed direction (e.g. the magnetic field) [21, 22]. Therefore, in liquids and gases, where the molecules undergo thermal motion, the spin interactions vary rapidly in time and the effective Hamiltonian is a time-average over this motion [16, 17]. For random molecular tumbling, the averaging only leaves the trace of the interaction tensors in the Hamiltonian. Therefore, dipolar and quadrupolar interactions, having traceless tensors, vanish, whereas chemical shielding and indirect coupling remain as apparent scalar interactions. This is the usual high-resolution condition, where anisotropic couplings do not show up in the NMR spectrum; spin relaxation effects caused by anisotropic spin interactions will nevertheless often be manifest. Whenever the tumbling is not completely random due to some aligning influence, the averaging will retain some of the tensorial nature of the spin interactions in the Hamiltonian. Hence anisotropic couplings are partly preserved and will affect the spectrum.

The effects of molecular orientation on NMR spectra have been discussed in detail in a number of accounts [1, 7, 16, 17, 23, 24]. In this section a simplified treatment is given to clarify the ideas involved. Attention is thereby restricted to the dipolar and the quadrupolar couplings, since these two are the principal interactions for alignment caused by a magnetic field [3]. In the following an isolated rigid molecule is considered. In doing so, effects of intermolecular interactions and of molecular vibrations are ignored. At the end of this section it is indicated how molecular correlation influences the degree of alignment. Vibrational corrections to molecular properties are generally small [7] and completely negligible in the present situation.

The dipolar and the quadrupolar Hamiltonians can both be expressed as contracted tensor products [21, 22]

$$H = \mathrm{h}\mathbf{I}\cdot\mathbf{A}\cdot\mathbf{S} = \mathrm{h}\sum_{\alpha',\beta'} \mathrm{A}_{\alpha'\beta'}\mathrm{I}_{\alpha'}\mathrm{S}_{\beta'} \qquad (\alpha', \beta' \in \mathrm{x}', \mathrm{y}', \mathrm{z}') \tag{1}$$

Here the time-averaged interaction tensor $\mathbf{A}$ and the spin vectors $\mathbf{I}$ and $\mathbf{S}$ are specified with respect to a laboratory frame (x′, y′, z′) in which the interactions are measured. The z′-axis is along the static magnetic field $\mathbf{B}$ and is thus the axis of quantization. Planck's constant h makes up for the dimensions: $\mathbf{I}$ and $\mathbf{S}$ are dimensionless spin operators and **A** is given in frequency units (Hz). For magnetic field alignment effects, the interaction in Eq. (1) can be treated as a first-order perturbation on the Zeeman energy [1]. Applying this so-called high field approximation, together with the fact that for dipolar and quadrupolar interactions the tensor **A** is symmetrical and traceless, yields [22, 24]

$$H = \mathrm{hA}[\mathrm{I}_{z'}\mathrm{S}_{z'} - \tfrac{1}{4}(\mathrm{I}_{+'}\mathrm{S}_{-'} + \mathrm{I}_{-'}\mathrm{S}_{+'})] \tag{2}$$

Here $\mathrm{A} = \langle \mathrm{A}_{z'z'} \rangle_{\mathrm{B}}$ is the motionally averaged anisotropic coupling constant that can be observed in the spectrum; $\mathrm{A}_{z'z'}$ is the component of **A** along the magnetic field direction.

To interprete the quantity A, it is to be linked to molecular properties. For this purpose a molecular frame (x, y, z) fixed in the molecule is introduced. In this frame the components $\mathrm{A}_{\alpha\beta}$ ($\alpha, \beta \in \mathrm{x, y, z}$) of **A** are constant in time (rigid molecule). An axes transformation then relates the tensor element $\mathrm{A}_{z'z'}$ to the $\mathrm{A}_{\alpha\beta}$. The transformation coefficients can be given in terms of the polar angles (θ, ϕ) which the magnetic field (z'-axis) makes with the molecular axes; see Fig. 2a. By using the fact that $\mathrm{A}_{z'z'}$ transforms as $(z')^2$ it can readily be shown that [3]

$$\begin{aligned} H = \langle \mathrm{A}_{z'z'} \rangle_{\mathrm{B}} = \mathrm{A}_{zz}\langle \tfrac{3}{2}\cos^2\theta - \tfrac{1}{2} \rangle_{\mathrm{B}} &+ \tfrac{1}{2}(\mathrm{A}_{xx} - \mathrm{A}_{yy})\langle \sin^2\theta \cos 2\phi \rangle_{\mathrm{B}} \\ &+ \mathrm{A}_{xy}\langle \sin^2\theta \sin 2\phi \rangle_{\mathrm{B}} + \mathrm{A}_{xz}\langle \sin 2\theta \cos\phi \rangle_{\mathrm{B}} \\ &+ \mathrm{A}_{yz}\langle \sin 2\theta \sin\phi \rangle_{\mathrm{B}} \end{aligned} \tag{3}$$

Since the molecular tumbling does not influence the $\mathrm{A}_{\alpha\beta}$, but only changes the angles θ and ϕ, the averaging on the r.h.s. of Eq. (3) is restricted to the angular functions $\mathrm{F}(\theta, \phi)$. The averages $\langle \mathrm{F}(\theta, \phi) \rangle_{\mathrm{B}}$ describe the molecular orientation and are commonly referred to as order parameters [16, 17]. For alignment by a magnetic field, the orienting mechanism is known [25]: the external magnetic field **B** interacts with the anisotropic magnetic susceptibility χ of the molecule. Therefore, using statistical mechanics (canonical ensemble), the order parameters can be

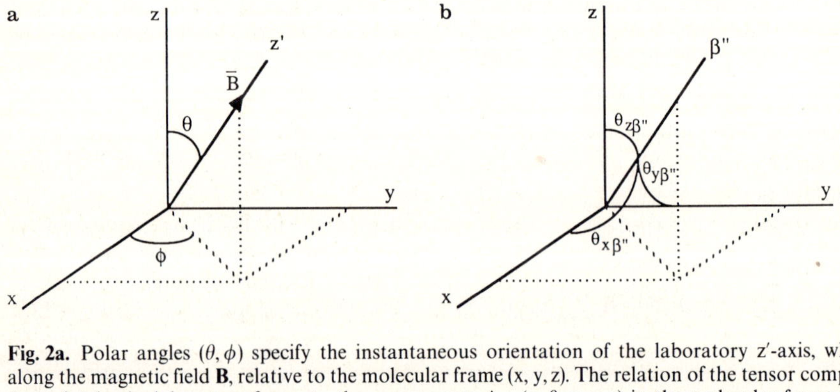

Fig. 2a. Polar angles (θ, ϕ) specify the instantaneous orientation of the laboratory z'-axis, which is along the magnetic field **B**, relative to the molecular frame (x, y, z). The relation of the tensor component $\mathrm{A}_{z'z'}$ of **A** in the laboratory frame to the components $\mathrm{A}_{\alpha\beta}$ ($\alpha, \beta \in \mathrm{x, y, z}$) in the molecular frame can be given in terms of θ and ϕ (see Eq. (3)). **(b)** The local β''-axis ($\beta'' \in \mathrm{x'', y'', z''}$) makes angles $\theta_{\alpha\beta''}$ ($\alpha \in \mathrm{x, y, z}$) with the molecular axes. The tensor components $\mathrm{A}_{\alpha\alpha}$ of **A** in the molecular frame can be related to the components $\mathrm{A}_{\beta''\beta''}$ in the local frame by means of the $\theta_{\alpha\beta''}$ (see Eq. (7))

expressed in χ, **B** and T, the absolute temperature [24, 26, 27]. It is advantageous to let the molecular frame coincide with the principal axes system of the susceptibility tensor, i.e. the frame in which χ is diagonal. For an isolated molecule the order parameters, up to first order in $(\chi B^2/\mu_0 kT)$ (high-temperature approximation), can then be calculated to be [1]

$$\langle \tfrac{3}{2}\cos^2\theta - \tfrac{1}{2} \rangle_B = \frac{\Delta\chi B^2}{15\,\mu_0 kT} \tag{4}$$

$$\langle \sin^2\theta \cos 2\phi \rangle_B = \frac{\delta\chi B^2}{15\,\mu_0 kT} \tag{5}$$

$$\langle \sin^2\theta \sin 2\phi \rangle_B = \langle \sin 2\theta \cos\phi \rangle_B = \langle \sin 2\theta \sin\phi \rangle_B = 0 \tag{6}$$

where μ_0 is the magnetic permeability of vacuum and k is the Boltzmann constant. The quantities $\Delta\chi = \chi_{zz} - \frac{1}{2}(\chi_{xx} + \chi_{yy})$ and $\delta\chi = \chi_{xx} - \chi_{yy}$ are the anisotropy and the asymmetry of the magnetic susceptibility respectively; the molecular axes are usually chosen such that $|\Delta\chi| \geqq |\delta\chi|$.

To complete the description the tensor elements $A_{\alpha\alpha}$ in Eq. (3) should be related to local directions within the molecule, like e.g. bonds. Therefore, a local frame (x'', y'', z'') is introduced, which is taken to be the principal axes system of the interaction tensor **A**; by convention the axes are chosen in such a way that $|A_{z''z''}| \geqq |A_{y''y''}| \geqq |A_{x''x''}|$. Since $A_{\alpha\alpha}$ transforms as $(\alpha)^2$, the transformation relation is [3]

$$A_{\alpha\alpha} = A_{z''z''}(\tfrac{3}{2}\cos^2\theta_{\alpha z''} - \tfrac{1}{2}) + \tfrac{1}{2}(A_{x''x''} - A_{y''y''})(\cos^2\theta_{\alpha x''} - \cos^2\theta_{\alpha y''}) \tag{7}$$

where it is assumed that **A** is traceless, as is the case for the dipolar and the quadrupolar interactions. The transformation coefficients are now expressed in terms of direction cosines: $(\cos\theta_{\alpha\beta''})$ is the cosine of the angle between the molecular α-axis and the local β''-axis (see Fig. 2b). These direction cosines are therefore determined entirely by the molecular geometry.

By combining Eqs. (3) to (7), the full expression for the anisotropic coupling constant A is obtained. The specific cases of dipolar and quadrupolar interactions will now be considered.

2.2 Dipolar Couplings

For two dipolarly coupled spins I and S, the tensor **A** in Eq. (1) represents the dipolar interaction tensor [22, 24]

$$\mathbf{D} = -\frac{\mu_0\gamma_I\gamma_S h}{8\pi^3 r^3}\left(\frac{3\,\mathbf{rr}}{2\,r^2} - \frac{1}{2}\mathbf{1}\right) \tag{8}$$

where γ_I and γ_S are the magnetogyric ratios of the spins and **r** is the internuclear distance vector. Therefore, within a local frame where the z''-axis is along **r**

$$D_{z''z''} = -2D_{x''x''} = -2D_{y''y''} = -\frac{\mu_0\gamma_I\gamma_S h}{8\pi^3 r^3} \tag{9}$$

The dipolar coupling constant D between spins I and S is then, combining Eqs. (3), (6), (7) and (9),

$$D = -\frac{\mu_0 \gamma_I \gamma_S h}{8\pi^3 r^3}[(\tfrac{3}{2}\cos^2\theta_{zz''} - \tfrac{1}{2})\langle\tfrac{3}{2}\cos^2\theta - \tfrac{1}{2}\rangle_B$$

$$+ \tfrac{3}{4}(\cos^2\theta_{xz''} - \cos^2\theta_{yz''})\langle\sin^2\theta\cos 2\phi\rangle_B] \tag{10}$$

Thus the dipolar Hamiltonian to be used for spectral interpretation is given by Eq. (2), with $A = D$.

For two coupled spins with $I = 1/2$ there are four spin energy levels, between which four allowed transitions are possible [1, 6]. This results in the well-known AB-quartet in the NMR spectrum. The frequency spacing $\Delta\nu_C$ of the left- and right-hand doublets in this pattern is given by

$$\Delta\nu_C = J + D \tag{11}$$

where J and D are the indirect and dipolar coupling constants. When the two spins are equivalent the A_2-case arises and the indirect coupling will not show up, leaving a dipolar effect. Now a doublet with a frequency difference $\Delta\nu_D$ between the two lines is expected (see Fig. 1)

$$\Delta\nu_D = \tfrac{3}{2}D \tag{12}$$

Equations (4), (5), (10), (11) and (12) fully describe the magnetic field induced dipolar couplings and the way in which they can affect the NMR spectrum.

2.3 Quadrupolar Couplings

The coupling of a nuclear quadrupole (only for spins with $I > 1/2$) to a surrounding electric field gradient is described by the quadrupolar interaction tensor [22, 24]

$$\mathbf{Q} = \frac{1}{2I(2I-1)}\left(\frac{eQ}{h}\right)\mathbf{V} \tag{13}$$

where eQ is the nuclear quadrupole moment and **V** is the electric field gradient tensor. Taking $A = Q$ and also, since only one spin is involved, $S = I$ in Eq. (2), the quadrupolar Hamiltonian becomes [22, 24]

$$H = hQ[\tfrac{3}{2}I_{z'}^2 - \tfrac{1}{2}I^2] \tag{14}$$

The anisotropic coupling constant Q follows from Eqs. (3), (6) and (13)

$$Q = \frac{1}{2I(2I-1)}\left(\frac{eQ}{h}\right)[V_{zz}\langle\tfrac{3}{2}\cos^2\theta - \tfrac{1}{2}\rangle_B$$

$$+ \tfrac{1}{2}(V_{xx} - V_{yy})\langle\sin^2\theta\cos 2\phi\rangle_B] \tag{15}$$

From Eq. (7), the relation between the components of the field gradient tensor in the molecular and local frames is usually expressed as

$$V_{\alpha\alpha} = eq[(\tfrac{3}{2}\cos^2\theta_{\alpha z''} - \tfrac{1}{2}) + \tfrac{1}{2}\eta(\cos^2\theta_{\alpha x''} - \cos^2\theta_{\alpha y''})] \tag{16}$$

where

$$eq = V_{z''z''} \tag{17}$$

and

$$\eta = \frac{V_{x''x''} - V_{y''y''}}{V_{z''z''}} \tag{18}$$

are the major component of the field gradient tensor and the so-called asymmetry parameter respectively.

For deuterium (^{2}H), with I = 1, the Zeeman interaction gives rise to three spin energy levels, between which two allowed transitions are possible. In the absence of quadrupolar interactions these two transitions are degenerate. Under the influence of the quadrupolar Hamiltonian in Eq. (12), however, the transition frequencies become unequal and a doublet appears in the spectrum with a frequency spacing [1, 6]

$$\Delta\nu_Q = 3Q = \frac{3}{2}\left(\frac{eQ}{h}\right)[V_{zz}\langle\tfrac{3}{2}\cos^2\theta - \tfrac{1}{2}\rangle_B$$
$$+ \tfrac{1}{2}(V_{xx} - V_{yy})\langle\sin^2\theta\cos 2\phi\rangle_B] \tag{19}$$

Equations (4), (5), (16) and (19) completely specify the magnetic field induced quadrupolar line splittings $\Delta\nu_Q$ in a deuterium NMR spectrum. The product (e^2qQ/h) that appears in $\Delta\nu_Q$ is often called the "quadrupole coupling constant".

2.4 Effects of Intermolecular Interactions

In the foregoing, the molecules were supposed to be isolated or non-interacting. For dense systems such as liquids, however, this assumption cannot be expected to hold. In fact, it is well known that molecules in liquids are in constant interaction with a large number of neighbouring molecules [24, 26–28]. Fortunately, most of the expressions given, remain valid when intermolecular interactions are taken into account: thus Eqs. (10), (11), (12) (dipolar case) and (16), (19) (quadrupolar case) will not be affected. It is in the calculation of the order parameters, where the differences arise. Equations (4), (5), (6) are strictly correct only for independent molecules. The presence of correlations between the molecules gives rise to extra contributions to the order parameters. For a liquid of N identical molecules with effective axial symmetry (only one order parameter differs from zero) it can be calculated that [27]

$$\langle\tfrac{3}{2}\cos^2\theta - \tfrac{1}{2}\rangle_B = g_2\frac{\Delta\chi B^2}{15\mu_0 kT} \tag{20}$$

where

$$g_2 = 1 + (N-1)\langle\tfrac{3}{2}\cos^2\theta_{ij} - \tfrac{1}{2}\rangle_0 \tag{21}$$

Here θ_{ij} is the angle between the molecular z-axes of molecules i and j, and the subscript 0 indicates an isotropic average, i.e. in the absence of the aligning magnetic field. In Eq. (20), $\Delta\chi$ is the magnetic susceptibility anisotropy of an isolated molecule and g_2 corrects for the correlation between the molecules. Note that interactions between induced magnetic moments give negligible contributions to the intermolecular potential; in this respect, magnetic fields are "cleaner" than electric fields where the description is complicated by internal fields. Since g_2 depends on the average over all the possible relative positions and orientations of the molecules, it is referred to as "angular correlation factor".

In the absence of angular correlation $g_2 = 1$. Equation (20) then allows the determination of molecular $\Delta\chi$-values from a measurement of the alignment via anisotropic couplings in the NMR spectrum. This limit ($g_2 = 1$) is in practice approximated by using a low concentration of solute molecules (small N) in a non-aligning solvent. In this way interactions between solute molecules are strongly reduced and alignment transfer from solvent to solute molecules is, in principle, absent. The concentrations at which deviations of g_2 from one become apparent, depend on the compound under study: strong correlations give rise to a large value of the average $\langle \frac{3}{2}\cos^2\theta_{ij} - \frac{1}{2}\rangle_0$. Concentration dependent measurements can be helpful in this respect.

The appearance of g_2 in the expressions for the order parameters might be seen as a nuisance. On the other hand, it opens the way to study intermolecular interactions by NMR alignment experiments [18–20]. In particular, temperature dependent studies appear to have a high potential for investigating angular correlation effects in liquids.

3 Analysis of Spectra

3.1 Experimental Aspects

In the previous section it was established how magnetic field alignment influences NMR spectra: the resonance lines are split or displaced due to the presence of residual anisotropic couplings. The line splittings $\Delta\nu$ (Eqs. (11), (12) and (19)) depend on a nuclear coupling constant ($(-\mu_0\gamma_I\gamma_S h/8\pi^3 r^3)$ for dipolar couplings; (e^2qQ/h) and η for quadrupolar couplings), the structure of the molecule and, via the order parameters, on the magnetic susceptibility tensor, the square of the magnetic field strength and on the absolute temperature (Eqs. (3)–(7)).

The experimental observation of these effects is straightforward, using a standard high-field NMR spectrometer. Obviously, line widths should be as small as possible. Therefore some care should be taken in preparing the sample. Thus high precision NMR tubes and clear solutions (without precipitate) should be used for an optimal field homogeneity, samples should be degassed to eliminate the line broadening due to paramagnetic oxygen, and a suitable solvent should be selected. The choice of solvent [9] is usually a compromise between requirements such as solubility of the solute, inertness (e.g. no specific solute-solvent interactions), viscosity and lock/shim-quality. For alignment studies, there is the additional constraint that the solvent itself be non-orienting (small $|\Delta\chi|$).

To obtain the dipolar coupling between two non-equivalent coupled spins it is necessary to separate the J and D contributions in $\Delta\nu_C$ (Eq. (11)). Since the indirect coupling J is generally larger than the line width, the resonances are already separated. In principle, then, small super-imposed dipolar couplings D can be determined by measuring the magnetic field dependence of $\Delta\nu_C = J + D(B^2)$. This also gives the relative signs of J and D.

Dipolar effects for two equivalent coupled spins or quadrupolar effects for a $I = 1$ spin can only be obtained if the doublet spacing ($\Delta\nu_D$ and $\Delta\nu_Q$) is large enough to overcome the line width. This situation is generally not reached. In most compounds studied so far; the component lines of the doublet severely overlap and e.g. resolution enhancement techniques [29, 30] must be applied to separate them (see Fig. 1).

Owing to the factor $(\gamma_I\gamma_S/r^3)$ in D, dipolar couplings are expected to be especially important in groups like ^{13}C–H, ^{13}C–F and CH_2, where the nuclei are closely spaced and have large magnetic moments. At presently available fields, dipolar effects for diamagnetic compounds amount up to a few tenths of a Hertz. Thus in e.g. a CH_2-group the line width, which for $I = 1/2$ nuclei usually is determined by dipolar relaxation, should preferably be of the same order of magnitude. A small line width, however, is not always a guarantee that a splitting will show up. The observed spin could be indirectly coupled to other spins in the molecule. Thus small unresolved long-range couplings may obscure the dipolar splitting. In order to remove these disturbing effects, decoupling of the interfering spins could be applied. An alternative to decoupling is presented in Fig. 1, which shows the CH_2-doublet in the 500 MHz proton NMR spectrum of fluorene. In order to get rid of small indirect couplings with the aromatic protons, the rings have been deuterated, thereby reducing the interfering couplings by a factor of $\gamma_H/\gamma_D \cong 6.5$. It is not always possible for spin systems of more than two spins to act in such a way that spectral interpretation in terms of simple relations like (11) and (12) is possible. Numerical simulation methods could then be used to extract the NMR parameters from the spectra.

For the quadrupolar case, the line width (LW) is governed by quadrupolar relaxation and $LW \propto (e^2qQ/h)^2$ [21, 22]. Since $\Delta\nu_Q \propto (e^2qQ/h)$ (Eqs. (16), (19)) a small quadrupole coupling constant is expected to be advantageous for the detection of the quadrupolar effect. Indeed, quadrupolar splittings for deuterium ($(e^2qQ/h) \cong 0.2$ MHz [31, 32]) have been observed for many compounds [1]; for nitrogen (usually $(e^2qQ/h) \geqq 1$ MHz [31]) lines are already too broad and no splittings can be observed, even in a field of 14.57 T (^{1}H: 620 MHz).

At currently available fields, deuterium quadrupolar effects for diamagnetic compounds range from less than 0.1 Hz to a few Hertz. The bigger effects arise for large molecules, having a large susceptibility anisotropy $\Delta\chi$. For these molecules, line widths are usually such that an overlapping doublet occurs in the spectrum (see Fig. 6). In general, both the line width and $\Delta\chi$ decrease when the size of the molecule is reduced. As a result many compounds of different sizes can be studied and the appearance of their doublets, determined by the ratio of $\Delta\nu_Q$ to LW, will be very much the same.

When considering perdeuteriated compounds some care should be exercised.

Indirect ^2H–^2H couplings might interfere with the quadrupolar couplings and mask the doublet splittings [9, 33]. The remedy in that case is to apply deuterium-decoupling or to study mono-deuterated compounds. Sometimes the indirect ^2H–^2H couplings are not manifested. This situation occurs for (usually larger) molecules, where the deuterium relaxation rates are large compared with the indirect couplings. The interfering couplings are then partially or completely averaged out; this is known as self-decoupling. Effects of indirect couplings of the deuterons with other nuclei, like protons, can of course also be eliminated by decoupling. Whenever it is impossible to obtain a pure doublet and coupling constants have to be determined from a complex spectral pattern, numerical simulation is again the method of choice for analysis.

In general the absolute signs of coupling constants cannot be obtained from an analysis of the NMR spectrum. The signs of the dipolar and quadrupolar couplings, however, are governed by their theoretical expressions. Therefore these signs can be inferred if sufficient information is available. For example, physical arguments concerning the susceptibility can give the signs of the order parameters in the equations.

3.2 Determination of Line Splittings

Once a good quality spectrum, or rather a free induction decay (FID), is recorded, the question arises how to obtain reliable quantitative frequency splittings $\Delta\nu$ from it. When the resonance lines are fully separated (no overlap), the situation is very straightforward. Thus the splitting $\Delta\nu_C$ for two non-equivalent spins can often be found directly from the measured line positions. More generally, however, the spacing $\Delta\nu_D$ or $\Delta\nu_Q$ between the overlapping lines of a doublet has to be determined. This could be accomplished by fitting some predefined function such as two shifted Lorentzian lines to the shape of the doublet, treating their positions, widths and heights as fit-parameters. More commonly, however, manipulations of the FID in the time-domain, prior to Fourier transformation, are used to extract the desired information.

In resolution enhancement techniques, the FID is multiplied by a suitable weighting function, so that after Fourier transformation reduced line widths result. A widely used method, available in commercial NMR software, is the Lorentzian–Guassian transformation [29]. It assumes a Lorentzian FID that decays in time according to $\exp(-t/T_2^*)$, with T_2^* the effective transverse relaxation time, and uses a weighting function of the form $\exp(At - Bt^2)$. It is generally recommended [34–36] that A is set equal to $1/T_2^*$, thereby annihilating the Lorentzian line shape, and that B, which gives rise to a Gaussian line shape, is adjusted for the desired resolution. When applied to doublets, the parameter B is chosen to make the component lines separated ("non-overlapping"; see Fig. 1). The splittings obtained in this way are correct within a few percent [6]. Additional errors arise if the original line shape is not completely Lorentzian. Thus e.g. inhomogeneity effects, impurity signals close to the doublet and interfering multiplicity of the lines should be avoided. Also, of course, the receiver circuit of the spectrometer should produce a good Lorentzian line shape.

The increase in resolution by this procedure is accompanied by a decrease of the signal-to-noise ratio (S/N). This poses a limit to the extent of enhancement attainable. In general, the maximum reduction of the line width amounts to about 35% [29]; after this the S/N decreases drastically and makes any quantitative determinations impossible. As a result, splittings for which $(LW/\Delta\nu) \leqq 2$ can be readily obtained.

An alternative choice of the parameters A and B has been suggested, that seems especially suited for the accurate determination of doublet line splittings [37]. Here A is set to cause over-enhancement ($A > 1/T_2^*$) so that wiggles of a periodic nature appear in the spectrum. The parameter B is then chosen to match these wiggles with the positions of the component lines of the doublet. The details of this approach will be discussed in a forthcoming paper [37]. The advantage here is that a much better S/N results in the enhanced spectrum. Consequently smaller splittings ($LW/\Delta\nu \leqq 3$) can be obtained and the accuracy is increased.

It should be noted that the Lorentzian–Gaussian transformation has inherently non-optimal S/N characteristics, i.e. the weighting function $\exp(At - Bt^2)$ is not a "matched filter". In the recently proposed TRAF (transform of reverse added FIDs) resolution enhancement method [38, 39] this matched filter concept is employed. As a result, the TRAF weighting function is capable of increasing resolution without seriously degrading S/N; in some cases both resolution and S/N may be improved simultaneously. In this respect, the TRAF method is superior to the Lorentzian–Gaussian enhancement technique and seems a promising alternative.

The above mentioned resolution enhancement techniques rely on assumptions about line shape. A new deconvolution technique which does not suffer from this,

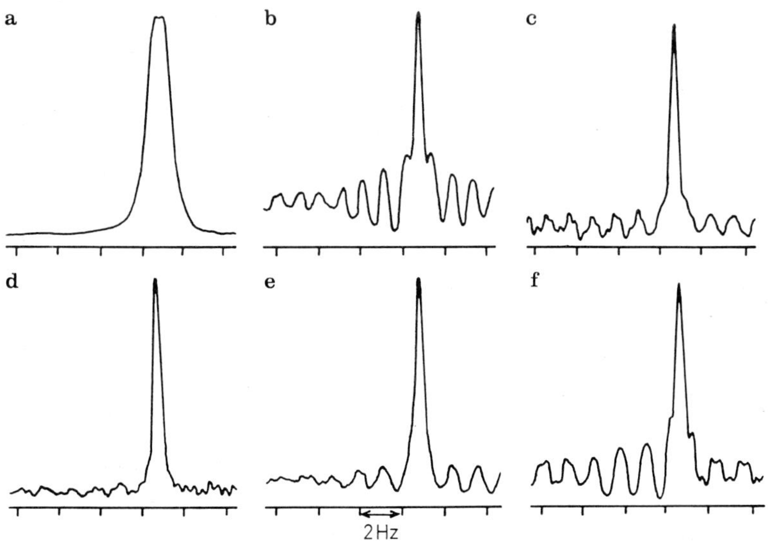

Fig. 3a–f. Illustration of doublet deconvolution. (**a**) 93.5 MHz (14.3 T) deuterium NMR spectrum of 2% $^2HCl_3$ in ether solution. Trial deconvolutions with (**b**) $\nu_S = 0.495$ Hz; (**c**) $\nu_S = 0.555$ Hz; (**d**) $\nu_S = 0.565$ Hz (matched); (**e**) $\nu_S = 0.595$ Hz; (**f**) $\nu_S = 0.635$ Hz. (Reproduced, with permission, from Ref. [40])

has lately been introduced [40]. It is based on the fact that any doublet, whatever the line shape, arises from an FID of the form S(t) cos ($\pi\Delta\nu$t), where $\Delta\nu$ is the doublet spacing. Multiplication of the FID from an arbitrary doublet by $1/\cos(\pi\nu_S t)$ will then convert it to the FID S(t) of a singlet of identical line shape, provided that $\nu_S = \Delta\nu$. For an incorrect choice of ν_S the line shape is not properly reconstructed, resulting in a broadened singlet of lower intensity and in modulation of the base-line. Accordingly an optimal ν_S exists which can be obtained with excellent precision (see Fig. 3). The only assumption in this method is that the doublet consists of two signals with identical line shapes. Extension to the deconvolution of arbitrary multiplets is readily made.

In general, therefore, the accuracy of dipolar and quadrupolar line splittings obtained from overlaping doublets by resolution enhancement or deconvolution techniques is about 2%.

4 Applications

4.1 Practical Consequences

From the preceding sections it is clear that high field NMR spectra may be influenced by orientation effects induced by the magnetic field. At the very high fields that will become available in the future (21.2 T (^{1}H: 900 MHz) and 23.5 T (^{1}H: 1 GHz) NMR spectrometers are already under development), these effects are expected to give substantial modifications of the experimental line positions for a large class of compounds (having a non-zero $|\Delta\chi|$). The way in which dipolar couplings affect the spectra of $I = 1/2$ spins (e.g. ^{1}H, ^{13}C, ^{19}F) [3] and quadrupolar couplings those of $I = 1$ spins (e.g. ^{2}H, ^{14}N) [9] has been treated in Sects. 2.2, 2.3. In addition, although their effects are much smaller, anisotropic chemical shifts and anisotropic indirect couplings may become manifest in the spectra [3, 7]. The implication is that spectrum interpretation is complicated, since the extra interactions have to be taken into consideration. The features of magnetic field alignment might therefore be seen as a hindrance in determining the usual NMR parameters (chemical shifts and indirect coupling constants). However, the effort it takes to obtain the anisotropic coupling constants is compensated for by the information that can be gained from them. The use of dipolar and quadrupolar line splittings to investigate molecular properties (e.g. magnetic susceptibilities, quadrupole coupling constants, structure) is discussed in the following sections. Many applications exploiting the occurence of magnetic field induced orientation effects have already appeared in the literature. Thus, for instance, measurements have been performed in the liquid phase as well as in the gas phase [15]; in solution a diverse range of compounds has been studied, covering small [20, 41] and large [4, 12] diamagnetic molecules (including ions [42, 43]) and paramagnetic systems [11, 44].

Before going on to the applications, there are a few more aspects of molecular orientation that need to be pointed out. One is concerned with delay times in multi-pulse experiments, like multiple quantum or echo spectroscopy. Strictly, as will be obvious from the foregoing, instead of $\tau=(4\,J)^{-1}$, a delay time $\tau=(4[J+D])^{-1}$ should be employed at high fields [9]. In order to assess the importance of the corrections on τ, the expressions for D in Sect. 2.2 can be very helpful.

Another important aspect is connected with the use of deuteriated locking agents. At high fields quadrupolar couplings may cause a splitting of the lock signal and thereby affect the lock stability and shimming procedures [9]. Thus frequency jumping of the lock might take place, leading to line broadening or doubling in the spectra or to t_1 noise in two-dimensional NMR experiments. For perdeuteriated compounds masking of this lock splitting due to interfering indirect ^2H–^2H couplings may occur and solve these problems. However, shimming and locking on the resulting unresolved multiplet is by far less sensitive than on a single sharp line. In general, therefore, deuteriated locking agents will become less useful at high magnetic fields; even at moderate fields, the problems mentioned may already play a role [9].

As a result, the requirements for a suitable locking substance are that it gives rise to narrow single lines and is non-aligning. The last condition is approached for molecules having a very small $|\Delta\chi|$ value; for these molecules indirect couplings between equivalent deuterons and quadrupolar splittings will hardly show up. In this respect perdeuteriated tetramethylsilane (TMS-d_{12}), having $\Delta\chi=0$, seems an ideal compound. When the locking substance and the solvent coincide, TMS-d_{12} has the additional qualities of inertness and low viscosity, but its poor dissolving capacity for many solutes is a disadvantage. Other suitable lock solvents are methylhalides, acetone, diethylether and alkanes.

4.2 Magnetic Susceptibilities

From a measurement of magnetic field induced dipolar or quadrupolar effects, the order parameters $\langle\frac{3}{2}\cos^2\theta-\frac{1}{2}\rangle_B$ and $\langle\sin^2\theta\cos 2\phi\rangle_B$ can be obtained; see Sect. 2, Eqs. (10) and (16), (19). Interpretation of these order parameters then allows the determination of molecular magnetic susceptibility anisotropies $\Delta\chi$ and asymmetries $\delta\chi$; see Eqs. (4), (5) and (20).

To find the order parameters from experimental line splittings some additional information is required. In the dipolar case the internuclear distance r has to be known and for the quadrupolar case the quadrupole coupling constant (e^2qQ/h) and the asymmetry parameter η have to be available. Also the structure of the molecule is needed for the direction cosines $(\cos\theta_{\alpha\beta''})$. The required data can usually be gathered from various literature sources. To obtain both order parameters, it is necessary to have the splittings of at least two nuclei in different geometrical positions in the molecule. Then (two) independent equations result from which the two unknowns can be solved.

The determination of molecular susceptibilities from the order parameters may be complicated by angular correlations. As was mentioned in Sect. 2.4, proper precautions should be taken to ensure that the influence of intermolecular interactions is negligibly small ($g_2 \cong 1$). Thus in the liquid phase, low solute concentrations and a non-aligning and non-interacting solvent should be employed. Suitable solvents are carbontetrachloride and TMS, having $\Delta\chi = 0$, but also alkanes, cycloalkanes, acetone and diethylether are appropriate.

Magnetic susceptibility anisotropies and asymmetries of many compounds have been measured by high-field NMR. A compilation of results for several aromatic systems and porphyrins is given in Ref. 1. Lately, a series of halomethanes [20, 41] and various compounds of biological interest [6] have been studied. In Table 1 the susceptibility data ($\Delta\chi, \delta\chi$) of newly investigated compounds are summarized. Since the values of $\Delta\chi$ and $\delta\chi$ depend on the particular choice of the molecular frame (x, y, z), the orientations of the axes have to be specified. Generally, z is taken along the principal molecular axis. Thus for molecules with effective axial symmetry (C_{3V} or higher) z is along the axis of highest symmetry and the choice of x and y is irrelevant ($\delta\chi = 0$). For the methylene halides CH_2X_2 in Table 1, z is along the C_{2V}-axis and y is in the plane of the two halide atoms. In aromatic systems, like the N-heterocycles in Table 1, the z-axis is chosen normal to the aromatic plane; convenient x and y directions usually follow from the molecular symmetry. For most planar aromatic molecules, the differences in the in-plane susceptibilities χ_{xx} and χ_{yy} are small [1] and can be ignored without introducing large errors (< 5%). The assumption $\delta\chi = 0$ was applied for some of the aromatics in Table 1, to find $\Delta\chi$ when only one experimental line splitting was available. Note that in this approximation the choice of x and y is again irrelevant.

The results in Table 1 are derived from dilute solutions, where the role of intermolecular interactions is expected to be small. Nevertheless, for some compounds with a strong tendency to aggregate, large errors may arise (viz. 1-methylimidazole, naphthimidazole, guanine [6]). Some experimental details are given in the table. For more information, both experimental and calculational (used geometries and quadrupole coupling constants), the original papers should be consulted. To obtain the accuracy of $\Delta\chi$ and $\delta\chi$, the error contributions of the quantities in the working equations for the line splittings (Eqs. (10) and (16), (19)) have to be added. In general: $\Delta\nu \cong 2\%$; $(e^2qQ/h) \cong 2$ to 4% (^2H); geometry(r or $\theta) \cong 1\%$; $T \cong 0.3\%$. This leads to an error in $\Delta\chi$ of about 7%; for $\delta\chi$ the absolute errors are the same as for $\Delta\chi$.

As can be seen from Table 1 only few literature data are available for the compounds listed. This reflects the inherent simplicity of the NMR method as compared to the other techniques [54, 55] for determining $\Delta\chi$. In the Cotton–Mouton effect [56] the optical birefringence due to magnetic field induced alignment is studied. It is a very sensitive technique that can be applied to liquids and gases, but it is experimentally difficult. Since bulk quantities are observed, concentration dependent measurements have to be performed. Furthermore, to obtain the magnetic susceptibilities, information is required about electric polarizabilities and internal fields. The Zeeman effect in microwave spectroscopy [57] can only be applied for gas phase studies. It is mainly suitable for small molecules, since the spectra become very complex as the size of the molecules increases. Also the molecules should possess a permanent dipole moment for interaction with the microwaves. The Krishnan critical angle method [58], finally,

Table 1. Magnetic susceptibility anisotropies, $\Delta\chi = \chi_{zz} - (\chi_{xx} + \chi_{yy})/2$, and asymmetries, $\delta\chi = \chi_{xx} - \chi_{yy}$, at roomtemperature (units: m^3/molecule; to convert the values in the table to cgs-units (emu/molecule or cm^3/molecule), divide by $4\pi \times 10^{-6}$). Details about the molecular axes definitions are given in the text

	NMR			Literature		
Compound	$10^{34}\,\Delta\chi$	$10^{34}\,\delta\chi$	Ref.	$10^{34}\,\Delta\chi$	$10^{34}\,\delta\chi$	Ref.
methyl chloride[a,1]	−1.57	0[e]	41	−1.66	0[e]	48[g]
methylene chloride[a,1]	0.00[f]	2.61	20, 41			
chloroform[a,2]	3.08	0[e]	20, 41	3.8; 2.4	0[e]	49; 50[h]
methyl bromide[a,1]	−1.73	0[e]	41	−1.77	0[e]	48[g]
methylene bromide[a,2]	0.00[f]	3.41	20, 41			
bromoform[a,2]	4.37	0[e]	20, 41	3.3	0[e]	50[h]
methyl iodide[a,1]	−2.00	0[e]	41	−2.29	0[e]	48[g]
methylene iodide[a,2]	0.00[f]	5.33	20, 41			
iodoform[a,2]	8.12	0[e]	20, 41			
1-methylimidazole[a,1]	−12	0.0[f]	6			
pyrimidine[a,3]	−12	−0.9	6			
benzimidazole[a,3,4]	−20.5	0.0[f]	6			
purine[a,3]	−19.5	0.0[f]	6			
adenine[a,4]	−19.1	0.0[f]	6			
guanine[b,c,5]	−18.3	0.0[f]	6			
naphthimidazole[b,c,3]	−51.5	0.0[f]	6			
porphyrin-quinone[b,d,j,6]	−121	0[f]	14			
DNA-decamer[b,d,k,7]	−152	0[f]	6			
styrene[a,2,8,9]	−14.5	0.0[f]	13			
ferrocene[a,2]	−16.0	0[e]	6	−12; −12	0[e]	51[h]; 52[i]
ruthenocene[a,2]	−9.8	0[e]	6	−6.7; −8.4	0[e]	51[h]; 52[i]
[3]-phenylene[a,l,1]	−19.4	0.0[f]	6			
triptycene[a,m,10]	24.0	0[e]	47			
tribenzotriptycene[a,m,10]	44.5	0[e]	47			

[a] ^{2}H measurements, B = 14.57 T (95.15 MHz); [b] ^{1}H field dependent measurements, highest field B = 14.57 T (620.2 MHz); [c] ^{13}C–^{1}H couplings; [d] ^{1}H–^{1}H couplings; [e] by symmetry; [f] assumed; [g] Zeeman effect in microwave spectroscopy (gas); [h] Cotton–Mouton effect; [i] Krishnan critical angle method (solid); [j] porphyrin-quinone cage molecule, see Fig. 7 (for details see Refs. 53, 54); [k] oligodeoxyribonucleotide helix in self-complimentary duplex form: d(C–C–A–A–G–C–T–T–G–G)$_2$; [l] see Fig. 4; [m] see Fig. 5. solvents: [1] diethyl-ether; [2] cyclohexane; [3] acetone; [4] ammonia; [5] D_2O/DCl; [6] methylene chloride; [7] D_2O/phosphate-buffer; [8] TMS/acetone; [9] cyclopentane/TMS; [10] carbondisulfide/acetone

is a solid state technique. Here problems may originate in the difficulty to relate the measured crystal susceptibilities to the molecular susceptibilities. An advantage of the NMR method is that in principle it can be applied to any compound. Unfortunately the effects are usually small and often deuteriated compounds have to be used (the quadrupolar effects are bigger than the dipolar effects, since the interaction is stronger). Also very strong magnetic fields are necessary

The NMR data generally agree within experimental error with susceptibility values obtained from microwave spectroscopy and from solid-state measurements. Differences between NMR and Cotton–Mouton results are often larger and may arise from the complexity of this last technique [18, 19]. The discrepancy in Table 1 between the $\Delta\chi$ values of methyl iodide for the NMR and microwave methods is probably due to resolution problems in the microwave spectrum for this particular case [41, 47].

4.3 Angular Correlation

The angular correlation factor g_2 accounts for the contribution of intermolecular interactions to the order parameters. In expressions like Eq. (20) it is assumed that the intermolecular interactions only affect the orientational behaviour of the molecules and not their intrinsic properties. Thus, for instance, the orientation of the principal frame of the magnetic susceptibility tensor should not be influenced by the molecular correlations. Note that this last requirement need not be fulfilled when besides angular correlation actual molecular complexes are formed (chemical reactions).

From Eq. (20) it follows that g_2 is the ratio of the order parameter obtained under the conditions of interest and under reference conditions where $g_2 \cong 1$. Consequently, g_2 may be determined from the concentration dependence of magnetic field induced NMR line splittings.

Measurements on substituted benzenes performed in this way give g_2 factors that agree quite well with results from depolarized Rayleigh scattering and the Cotton–Mouton effect [1, 18, 19]. Recently, the NMR method was applied to chloroform and bromoform [20]. The g_2 factors of the pure compounds are compared with literature values in Table 2. The NMR and Cotton–Mouton data agree satisfactorily. However, there is an enormous discrepancy with the other optical techniques. Since the NMR method is simple and straightforward, the NMR g_2 values are believed to be reasonably accurate. The optical techniques are experimentally difficult and require knowledge of complex quantities such as polarizabilities and internal field factors. Nevertheless, the origin of the large deviations is not easily assessed and remains unclear; further investigations by e.g. temperature dependent NMR alignment experiments are expected to give more information concerning angular correlation effects in the haloforms.

The interpretation of the g_2 factors is extremely complex. It relies, via the averages in expressions like Eq. (21), on the detailed form of the intermolecular interaction potentials in the system under investigation. Unfortunately, due to the complexity of intermolecular interactions in dense systems, these interaction potentials are not known accurately and empirical model potentials have to be used instead. Therefore, via the study of g_2 factors, information can be obtained about these model potentials. In this respect, as already mentioned, investigating the temperature dependence of g_2 via the NMR magnetic field alignment method seems very promising.

Table 2. Angular correlation factors g_2 for chloroform and bromoform. Comparison of results from the NMR magnetic field alignment method (NMR), the Cotton–Mouton effect (CM), depolarized Rayleigh scattering (RS) and Rayleigh/Raman line shape analysis (RR)

Compound	g_2(NMR)[a]	g_2(CM)[b]	g_2(RS)[c]	g_2(RR)[d]
chloroform	1.07	1.25	1.6	1.6
bromoform	1.13			2.3

[a]Ref. 20; [b]Ref. 48; [c]Ref. 59; [d]Ref. 60

4.4 Aromaticity

The cyclic delocalization of π-electrons in aromatic systems gives rise to an enhanced susceptibility component χ_{zz} perpendicular to the aromatic ring plane. As a result, the magnetic susceptibility anisotropy $\Delta\chi$ of an aromatic molecule is much larger than that of an analogous nonaromatic molecule where all electrons are localized in bonds. Thus the $\Delta\chi$ value of a compound should be indicative of its aromatic character [61, 62]. This principle makes the NMR alignment technique well suited to study aromaticity. For example, it was used to establish that [5]-metacyclophane, a strained bridge compound with a non-planar benzene ring, is aromatic [62]. A few other examples where the $\Delta\chi$ value provides interesting information concerning aromaticity can be found in Table 1.

The [3]-phenylene molecule [6] consists of three benzene rings and two cyclobutadiene rings (Fig. 4). However, its $\Delta\chi$ value (-19.4×10^{-34} m^3/molecule [6]) amounts to only 50% of the value for anthracene (-36.9×10^{-34} m^3/molecule [1]). This reduced aromaticity of [3]-phenylene as compared to anthracene is attributed to the presence of the four-membered rings. Generally, simple monocyclic systems with $4n + 2$ π-electrons, where n is an integer, are aromatic, i.e. π-electron delocalization leads to a stabilization of the compound (Hückel rule); for systems with $4n$ π-electrons, on the other hand, a destabilization may occur and the term anti-aromatic is used [63]. Clearly, the cyclobutadiene parts of [3]-phenylene are of the 4n-type and they oppose the cyclic delocalization of the π-electrons brought about by the benzene parts ($4n + 2$ type), thereby lowering the aromatic character of the molecule [6].

For the benzotriptycenes in Fig. 5 [46], different interpretations of the observed proton chemical shifts led to conflicting conclusions concerning the aromaticity of these compounds. In one paper [64], it was asserted that the aromaticity is reduced by a factor of approx. 2 as compared with planar aromatic systems, whereas in the other work [65] the aromatic character appeared to be normal. As a more direct check of the aromaticity of these compounds, their $\Delta\chi$ values

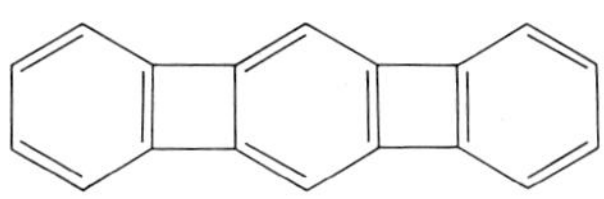

Fig. 4. [3]-phenylene

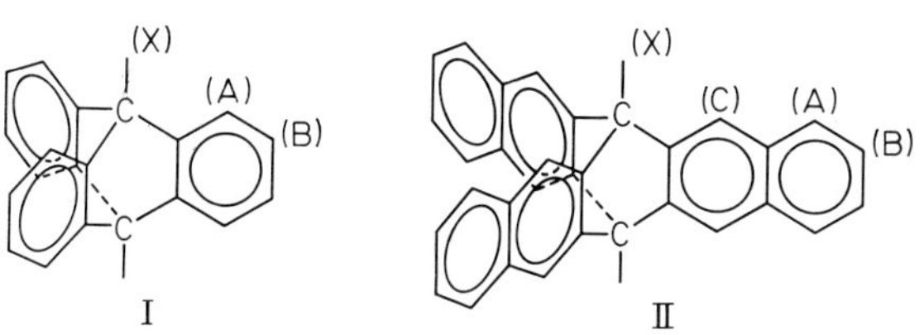

I II

Fig. 5. (I) Triptycene; (II) Tribenzotriptycene. (Reproduced, with permission, from Ref. [46])

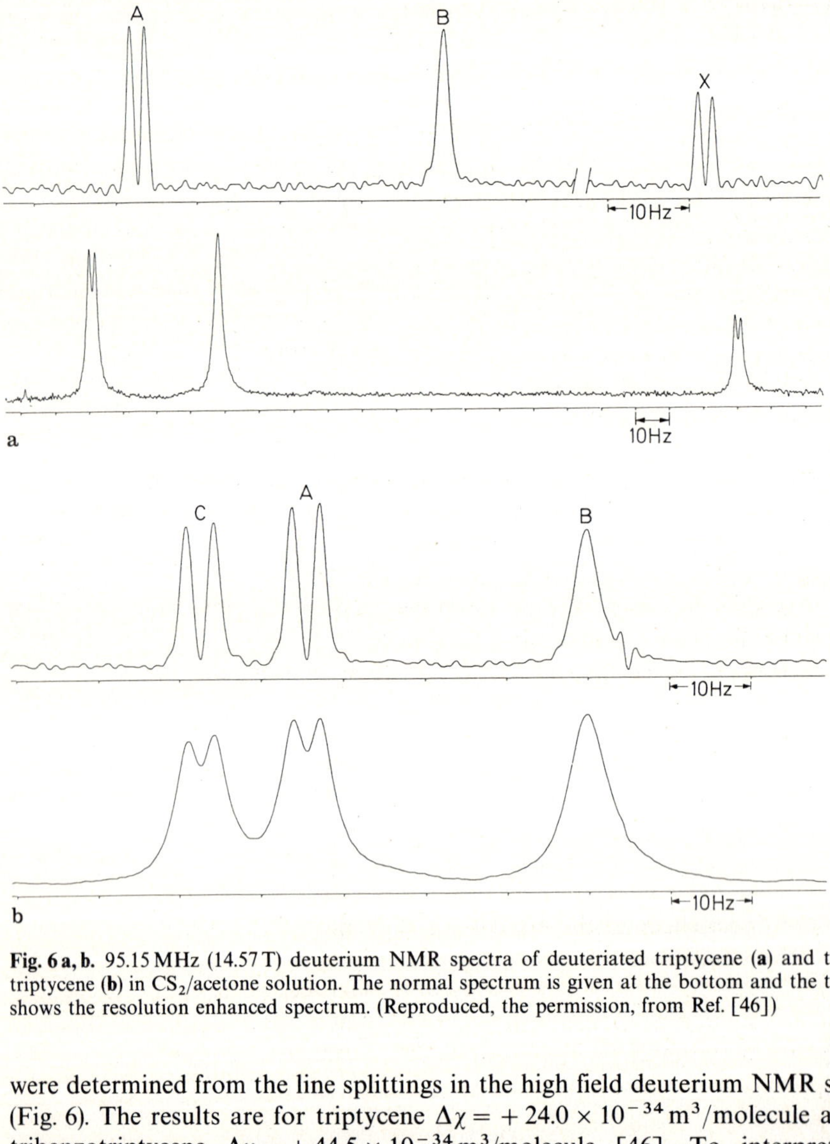

Fig. 6 a, b. 95.15 MHz (14.57 T) deuterium NMR spectra of deuteriated triptycene (**a**) and tribenzotriptycene (**b**) in CS_2/acetone solution. The normal spectrum is given at the bottom and the top trace shows the resolution enhanced spectrum. (Reproduced, the permission, from Ref. [46])

were determined from the line splittings in the high field deuterium NMR spectra (Fig. 6). The results are for triptycene $\Delta\chi = +24.0 \times 10^{-34}\,\mathrm{m}^3$/molecule and for tribenzotriptycene $\Delta\chi = +44.5 \times 10^{-34}\,\mathrm{m}^3$/molecule [46]. To interprete the obtained $\Delta\chi$ values of the benzotriptycenes, it seems proper to relate them to the susceptibilities of suitable fragments within the molecules. Thus, considering triptycene and tribenzotriptycene to be composed of independent benzene and naphthalene parts respectively, it can be shown that

$$\Delta\chi(\text{model}) = -\tfrac{3}{2}\Delta\chi(\text{f}) + \tfrac{9}{4}\delta\chi(\text{f}) + \Delta\chi(\text{corr}) \qquad (22)$$

where $\Delta\chi(\text{f})$ and $\delta\chi(\text{f})$ are the susceptibility anisotropy and asymmetry of the fragment molecules. For benzene $\Delta\chi(\text{b}) = -12.7 \times 10^{-34}\,\mathrm{m}^3$/molecule and $\delta\chi(\text{b}) = 0$ [18], and for naphthalene $\Delta\chi(\text{n}) = -25.0 \times 10^{-34}\,\mathrm{m}^3$/molecule and

$\delta\chi(\mathrm{n}) = +0.6 \times 10^{-34}\,\mathrm{m}^3$/molecule [1]. The correction term accounts for the cage in the benzotriptycenes; from bond susceptibilities [66] and the molecular geometry $\Delta\chi(\mathrm{corr}) = +1.8 \times 10^{-34}\,\mathrm{m}^3$/molecule [46]. Therefore it follows that for triptycene $\Delta\chi(\mathrm{model}) = +20.8 \times 10^{-34}\,\mathrm{m}^3$/molecule and for tribenzotriptycene $\Delta\chi(\mathrm{model}) = +40.6 \times 10^{-34}\,\mathrm{m}^3$/molecule. The values of $\Delta\chi(\mathrm{model})$ are 10 – 15% lower than the experimental $\Delta\chi$ values, showing that the benzotriptycenes are about as aromatic as planar aromatic systems [46].

4.5 Molecular Conformations

The working equations for the experimental line splittings contain the molecular geometry via the direction cosines ($\cos\theta_{\alpha\beta''}$): the angles $\theta_{\alpha\beta''}$ relate local directions, like for instance bonds, to the molecular frame. Therefore if all other quantities in the expressions are known, structure information can be obtained. Usually in this kind of application, the line splittings of several geometrically different nuclei within a molecule are combined to eliminate the order parameters. This has the advantage that the results are not influenced by angular correlation effects and do not rely on susceptibilities from other experiments.

For instance, in the study of a DNA-decamer d(C–C–A–A–G–C–T–T–G–G)$_2$ in solution [6] (see also Table 1), the angle between the planes of a central and a terminal base was inferred from two experimental dipolar couplings in the respective bases. The dipolar effects were observed for a certain proton pair (C5H, C6H) in two different cytosine residues: 1cyt (terminal) and 6cyt (center). Under the reasonable assumption that the oligodeoxyribonucleotide has axial symmetry about the helical axis ($\delta\chi = 0$), the ratio of the dipolar couplings is from Eq. (10)

$$\frac{\mathrm{D(1cyt)}}{\mathrm{D(6cyt)}} = \frac{(\frac{3}{2}\cos^2\theta_{zz''}(\mathrm{1cyt}) - \frac{1}{2})}{(\frac{3}{2}\cos^2\theta_{zz''}(\mathrm{6cyt}) - \frac{1}{2})} \tag{23}$$

where $\theta_{zz''}$ is the angle between the helical axis (z) and the (C5H, C6H) interproton vector. The measured coupling constants are $\mathrm{D(1cyt)} = -0.21$ Hz and $\mathrm{D(6cyt)} = -0.26$ Hz at B = 14.57 T (^{1}H: 620.2 MHz). The plane of the central base will be almost normal to the helical axis, so that $\theta_{zz''}(\mathrm{6cyt}) \cong 90°$. It therefore follows that $\theta_{zz''}(\mathrm{1cyt}) \cong 75°$ and that the angle between the two cytosine base planes is at least 15°, indicating that the tight double helix structure is loosening up at the ends of the strand.

This example illustrates the principle of the procedure: measurement of anisotropic interactions permits geometric constraints to be placed on the molecular structure. Recently, the difficult problem of determining the 3-dimensional solution conformation of a complex cage molecule by high field ^{1}H NMR was tackled [14]. The case molecule contains a porphyrin and a quinone held in a cofacial orientation by four identical bridging groups (see Fig. 7). When the bridges are extended, the molecule has C_{2v} symmetry and the quinone is at its maximum distance from the porphyrin ring. From the NMR measurements, the solution conformation was found to be a twisted structure, in which the bridges are wound into a spiral array about the C_2-axis, thereby shortening the porphyrin–

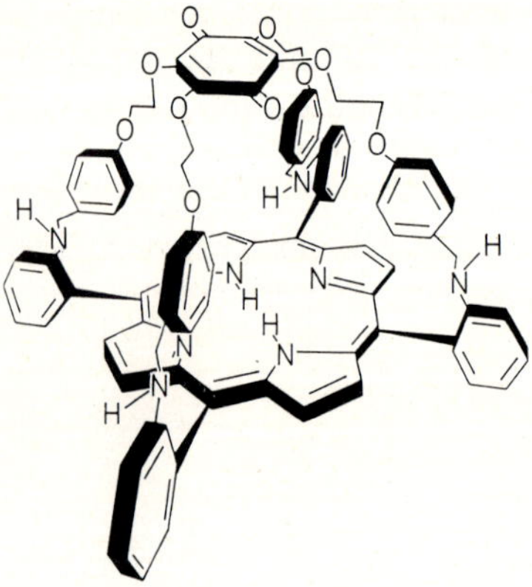

Fig. 7. Porphyrin-quinone cage molecule. (Reproduced, with permission, from Ref. [14])

quinone interplanar distance. Two twisted conformers are possible, differing in the distance between the porphyrin and quinone rings. The porphyrin–quinone cage molecule serves as a model system for the study of photosynthetic processes and as such knowledge of the 3-dimensional solution structure is essential. The inferred conformations are discussed in detail in the original paper [14].

So far the molecules were treated as being rigid and all molecular properties were considered to be static. For flexible molecules, however, depending on the type of motion that takes place, some properties will be modulated. For example, the geometry factors $(\frac{3}{2}\cos^2\theta_{\alpha z''} - \frac{1}{2})$ and $(\cos^2\theta_{\alpha x''} - \cos^2\theta_{\alpha y''})$ in Eqs. (10) and (16) and also the molecular susceptibilities may become averages over the internal motions. The experimental line splittings then only permit these averages to be obtained. Note that in this situation no straightforward structural information is available: an interpretation of the averages enables a characterization of the dynamics of the system. Eventually only an average or most probable structure may be deduced from the dynamics. This will be illustrated for the internal rotation in the styrene molecule (see Fig. 8a, b).

Styrene (deuteriated) was studied by high field ^{2}H NMR [13]. Due to the rotation of the vinyl group relative to the benzene ring, some of the deuterons are not fixed in the molecular frame and this is reflected in their quadrupolar line

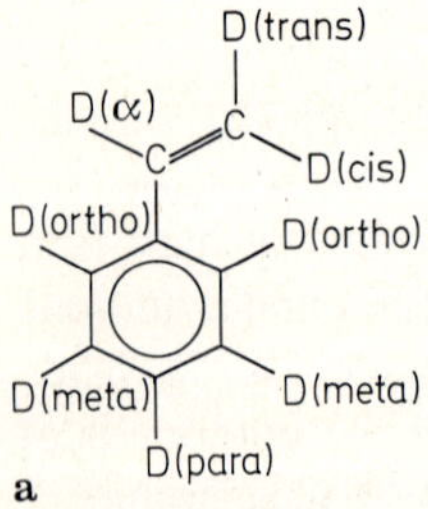

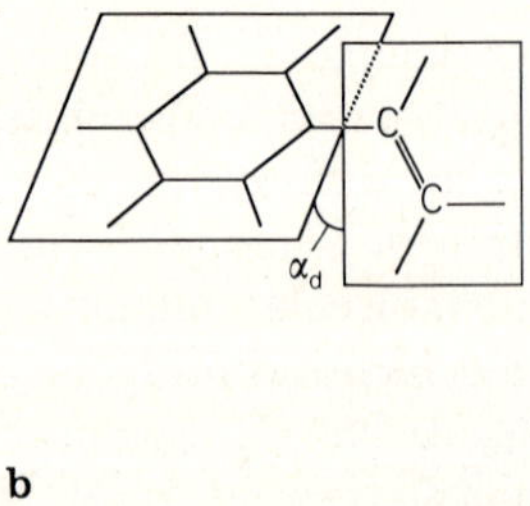

Fig. 8a. Styrene. **b** Internal rotation in styrene: α_d is the instantaneous dihedral angle between the phenyl and vinyl planes

splittings. To simplify the analysis of these splittings some approximations are introduced. Firstly, since the vinyl group contributes at most approx. 10% to the molecular susceptibility, it is assumed that the molecular z-axis remains perpendicular to the phenyl-group. From the observation that the line splittings $\Delta\nu_Q$(arom) for all aromatic deuterons are identical within experimental error it then follows that $\delta\chi \cong 0$ (the $\Delta\chi$ value obtained from $\Delta\nu_Q$(arom) is given in Table 1). As a further approximation the quadrupole coupling constants for the vinyl group deuterons are taken to be equal and their asymmetry parameters η are neglected. A final assumption concerns the vinyl group geometry: the (C–D) bond in the *trans*-position is taken parallel to the rotational axis (molecular x-axis), so that its line splitting $\Delta\nu_Q(trans)$ should not be influenced by the internal rotation. From the foregoing and Eqs. (16) and (19) it is expected that the ratio $\Delta\nu_Q$(arom)/$\Delta\nu_Q(trans)$ equals $[e^2qQ/h(\text{arom})]/[e^2qQ/h(trans)]$; this is confirmed by experiment, supporting the approach so far. Also the (C–D) bonds in the α and *cis*-positions are taken parallel and the line splittings for the α and *cis*-deuterons should therefore be identical. Their values, however, differ by an amount which is about the estimated error ($\pm 4\%$) and this is attributed to small differences in the bond angles or in the quadrupole coupling constants. For the calculation, the splitting $\Delta\nu_Q(cis)$ is used. Under the above assumptions, the ratio of $\Delta\nu_Q(cis)$ and $\Delta\nu_Q(trans)$ is from Eqs. (16) and (19)

$$\frac{\Delta\nu_Q(cis)}{\Delta\nu_Q(trans)} = \frac{\langle \frac{3}{2}\cos^2\theta_{zz''}(cis) - \frac{1}{2}\rangle}{\langle \frac{3}{2}\cos^2\theta_{zz''}(trans) - \frac{1}{2}\rangle} = \frac{(\frac{3}{2}\langle \sin^2\alpha_d\rangle \sin^2\theta_{xz''}(cis) - \frac{1}{2})}{(\frac{3}{2}\langle \sin^2\alpha_d\rangle \sin^2\theta_{xz''}(trans) - \frac{1}{2})} \tag{24}$$

where $\theta_{zz''}$ is the angle between the molecular z-axis and the (C–D) bond (z″-axis). In the last term of Eq. (24) the (C–D) bonds are related to the internal rotation axis (x) by expressing $\theta_{zz''}$ in the more convenient angles α_d and $\theta_{xz''}$: α_d is the dihedral angle formed by the phenyl and vinyl planes (see Fig. 8b) and $\theta_{xz''}$ is the angle between the rotational axis and the (C–D) bond. From the assumed vinyl group geometry $\theta_{xz''}(trans) = 0°$ and $\theta_{xz''}(cis) \cong 120°$. The measured line splittings are $\Delta\nu_Q(trans) = 0.565$ Hz and $\Delta\nu_Q(cis) = 0.443$ Hz at $B = 14.57$ T, so that from Eq. (24) $\langle \sin^2\alpha_d\rangle = 0.081$.

To characterize the internal rotation in styrene, use is made of the statistical mechanical (canonical ensemble) expression for the average

$$\langle \sin^2\alpha_d\rangle = \frac{\int \sin^2\alpha_d \exp(-V(\alpha_d)/kT)\, d\alpha_d}{\int \exp(-V(\alpha_d)/kT)\, d\alpha_d} \tag{25}$$

For styrene the internal rotation is assumed to be largely governed by a two-fold barrier with the planar form being the most stable [13]. Therefore a suitable potential energy function will be of the form

$$V(\alpha_d) = V_0 \sin^2\alpha_d \tag{26}$$

Numerical evaluation of the average in Eq. (25) for a range of V_0 values then allows a determination of the energy barrier to internal rotation: $V_0 = 17$ kJ/mol. In principle the model potential from Eq. (26) can now be used to evaluate the averages of any other quantities that depend on α_d. For example, the root-mean-square dihedral angle becomes $\alpha_d(\mathrm{RMS}) = \langle \alpha_d^2 \rangle^{1/2} = 16.5°$. Note, however, that this result does not specify an actual molecular structure, the styrene molecule being a dynamic entity. It merely reflects the mean deviation of α_d from the most stable geometry, which for styrene is the planar conformation [13].

4.6 Quadrupole Coupling Constants

The NMR magnetic field alignment method can also be used to obtain information on quadrupole coupling constants (e^2qQ/h), the procedure being somewhat analogous to that for the conformation studies. Thus the line splittings (quadrupolar as well as dipolar) of several nuclei within a molecule may be combined to eliminate the order parameters. When the molecular geometry is available, the nuclear coupling constants can be compared. An example of this was already encountered in the previous section where for styrene the splitting ratio $\Delta\nu_Q(\mathrm{arom})/\Delta\nu_Q(\mathit{trans})$ was found to be equal to the ratio of quadrupole coupling constants $[e^2qQ/h(\mathrm{arom})]/[e^2qQ/h(\mathit{trans})]$.

Similarly in deuteriated triptycene [47] (Fig. 5) it is expected that $\Delta\nu_Q(X)/\Delta\nu_Q(A)$ equals $[e^2qQ/h(X)]/[e^2qQ/h(A)]$. For deuteron A the usual quadrupole coupling constant for an aromatic deuteron is assumed: $e^2qQ/h(A) \cong 186\,\mathrm{kHz}$. Since the measured splitting ratio is 0.98, it follows that $e^2qQ/h(X) \cong 182\,\mathrm{kHz}$, which differs from the value of approx. 165 kHz for a normal deuteron bonded to an sp^3-hybridized carbon. This deviation is attributed to the strain in the cage to which deuteron X is connected. An increase of $e^2qQ/h(X)$ is in accordance with the reduced symmetry around the sp^3-hybridized bridge-head carbon due the strain (loss of *p*-orbital character, towards sp^2).

In general, the above comparison procedure for obtaining quadrupole coupling constants by the NMR magnetic field alignment method is quite reliable, since it hardly involves any assumptions. Recently a discussion concerning the correctness of quadrupole coupling constants for the complete series of deuteriohalomethanes was given [41]. Based on experimental data from microwave spectroscopy, from low-temperature solid-state NQR and NMR, and from NMR on liquid crystal mixtures, preferred values for the quadrupole coupling constants were deduced. The approach is supported by results from high-field high-resolution deuterium NMR.

5 Concluding Remarks

In the preceding sections several aspects connected with molecular orientation in NMR caused by the magnetic field of the spectrometer have been considered. Practical consequences were pointed out (Sect. 4.1) to make NMR spectroscopists alert to these alignment effects, which even at currently available field strengths may be substantial. When recognized, the features of NMR magnetic field alignment can be used to advantage. Thus information has been obtained about molecular susceptibility anisotropies and asymmetries (Sect. 4.2), intermolecular interactions (Sect. 4.3), aromaticity (Sect. 4.4), solution conformations (Sect. 4.5) and quadrupole coupling constants (Sect. 4.6). The development of higher fields in NMR spectroscopy will facilitate these studies, since both the accuracy and the detection limit will increase. The power of the NMR magnetic field alignment method, as compared to other techniques, lies in its inherent simplicity: experimentally, it is not complex, it is widely applicable and usually the interpretation is straightforward, relying on only a few non-critical assumptions.

Although the same applications, as were given for diamagnetic molecules, can also be made for paramagnetic molecules, results for the latter are much sparser. Recently, high field deuterium NMR was invoked in an analysis of isotropic paramagnetic shifts induced by Lanthanide shift reagents (LSR) [1, 45]. The dipolar shift contributions for deuterons in pyridine-d_5 bound to LSR were evaluated from quadrupolar line splittings. From a comparison with the experimental shifts it was concluded that Fermi contact shifts do not contribute significantly.

Most of the quadrupolar nuclei are unsuitable for study by high-resolution NMR spectroscopy owing to severe line broadening caused by efficient quadrupolar relaxation. Deuterium with its small quadrupole coupling constant is, in this respect, a lucky exception, as is also exemplified by the applications in the previous section. To make the quadrupolar nuclei more accessible for high-resolution studies, there is great interest in the minimization of the quadrupolar relaxation, for example by lowering the solution viscosity. As such, supercritical fluids, whose viscosities can be up to 2 orders of magnitude less than those of typical fluids, have been suggested as solvents [67]. Indeed, it was demonstrated that solutes containing quadrupolar nuclei, when dissolved in supercritical fluid solvents undergo a drastic reduction of their NMR line widths [67, 68]. The potential of these solvents for use in NMR alignment experiments therefore seems promising: if sufficient line narrowing can be reached, extension of the method to quadrupolar nuclei like ^{14}N and ^{17}O might become possible.

In conclusion, it can be stated that the NMR magnetic field alignment method is a convenient technique for obtaining information on a diversity of applications. It is expected that the use of supercritical fluids in combination with the future availability of higher magnetic fields, will greatly enlarge the range of possibilities of the method.

Acknowledgments

The authors thank Drs. A. A. Bothner-By, J. Dadok, C. M. Skoglund and P. C. M. van Zijl (Pittsburgh, USA) for kindly providing their latest manuscripts. Mr. L. Huis (Amsterdam, The Netherlands) is thanked for stimulating discussions and for carefully reading the manuscript. Mr. E. T. J. Nibbering (Amsterdam) is acknowledged for preparing fluorene-d_8 and for recording the spectrum presented in Fig. 1. Most of the data reported, were obtained at the NMR Facility for Biomedical Studies (620 MHz spectrometer) in Pittsburgh, supported by NIH grant RR 00292. The cooperation between the laboratories in Pittsburgh and Amsterdam was greatly facilitated by a NATO travel grant.

References

1. Bastiaan EW, MacLean C, van Zijl PCM, Bothner-By AA (1987) in: Webb GA (ed). Academic, London, pp 35–77 (Annual reports on NMR spectroscopy, vol 19)
2. van Zijl PCM, Ruessink BH, Bulthuis, J, MacLean C (1984) Acc. Chem. Res. 17: 172
3. Bastiaan EW, Bulthuis J, MacLean C (1986) Magn. Reson. Chem. 24: 723
4. Gayathri C, Bothner-By AA, van Zijl PCM, MacLean C (1982) Chem. Phys. Lett. 87: 192
5. van Zijl PCM (1987) Trends Anal. Chem. 6: 23
6. Skoglund CM (1987) Molecular orientation studies using high field NMR. Thesis, Carnegie-Mellon University, Pittsburgh
7. Lounila J, Jokisaari J (1982) Prog. NMR Spectrosc. 15: 249
8. Lohman JAB, MacLean C (1978) Chem. Phys. 35: 269; Lohman JAB, MacLean C (1979) Chem. Phys. 43: 144
9. van Zijl PCM (1987) J. Magn. Reson. 75: 335
10. Anet FAL (1986) J. Am. Chem. Soc. 108: 1354
11. Bothner-By AA, Domaille PJ, Gayathri C (1981) J. Am. Chem. Soc. 103: 5602
12. Bothner-By AA, Gayathri C, van Zijl PCM, MacLean C, Lai JJ, Smith KM (1985) Magn. Reson. Chem. 23: 935
13. Facchine KL, Staley SW, van Zijl PCM, Mishra PK, Bothner-By AA (1988) J. Am. Chem. Soc. 110: 4900
14. Lisicki MA, Mishra PK, Bothner-By AA, Lindsey JS (1988) J. Phys. Chem. 92: 3400
15. van Zijl PCM, MacLean C, Skoglund CM, Bothner-By AA (1985) J. Magn. Reson. 65: 316
16. Saupe A (1964) Z. Naturforsch. 19A: 161
17. Snyder LC (1965) J. Chem. Phys. 43: 4041
18. van Zijl PCM, MacLean C, Bothner-By AA (1985) J. Chem. Phys. 83: 4410
19. van Zijl PCM, MacLean C, Bothner-By AA (1985) J. Chem. Phys. 83: 4978
20. Bothner-By AA, Dadok J, Mishra PK, van Zijl PCM (1987) J. Am. Chem. Soc. 109: 4180
21. Abragam A (1985) The principles of nuclear magnetism. Clarendon, Oxford
22. Slichter CP (1980) Principles of magnetic resonance. Springer, Berlin Heidelberg New York
23. Diehl P, Khetrapal CL (1969) in: Diehl P, Fluck E, Kosfeld R (eds). Springer, Berlin Heidelberg New York, pp 1–95 (NMR basic principles and progress, vol 1)
24. Hilbers CW, MacLean C (1972) in: Diehl P, Fluck E, Kosfeld R (eds). Springer, Berlin Heidelberg New York, pp 1–52 (NMR basic principles and progress, vol 7)
25. van Vleck JH (1932) The theory of electric and magnetic susceptibilities. Oxford University Press, Oxford
26. Hilbers CW, MacLean C, Mandel M (1971) Physica 51: 246
27. Buckingham AD (1967) Discuss. Faraday Soc. 43: 205
28. Kirkwood JG (1939) J. Chem. Phys. 7: 919
29. Ernst RR (1966) in: Waugh JS (ed). Academic, New York, pp 1–135 (Advances in magnetic resonance, vol 2)
30. Ernst RR, Bodenhausen G, Wokaun A (1987) Principles of NMR in one and two dimensions. Clarendon, Oxford
31. Loewenstein A (1983) in: Smith JAS (ed). Heyden, London, pp 53–82 (Advances in nuclear quadrupole resonance, vol 5)
32. Jacobsen JP, Pedersen EJ (1981) J. Magn. Reson. 44: 101
33. Bothner-By AA, Gayathri C, van Zijl PCM, MacLean C (1984) J. Magn. Reson. 56: 456
34. Pearson GA (1987) J. Magn. Reson. 74: 541
35. Lindon JC, Ferrige AG (1980) Prog. NMR Spectrosc. 14: 27

36. Ferrige AG, Lindon JC (1978) J. Magn. Reson. 31: 337
37. Dadok J (1988) To be published.
38. Traficante DD, Nemeth GA (1987) J. Magn. Reson. 71: 237
39. Traficante DD, Ziessow D (1986) J. Magn. Reson. 66: 182
40. Bothner-By AA, Dadok J (1987) J. Magn. Reson. 72: 540
41. van Zijl PCM, Bothner-By AA (1988) J. Magn. Reson. 79: 439
42. van Zijl PCM, Velthorst NH, MacLean C (1983) Mol. Phys. 49: 315
43. van Zijl PCM, Kostermans GBM, MacLean C (1985) J. Am. Chem. Soc. 107: 2641
44. Domaille PJ (1980) J. Am. Chem. Soc. 102: 5392
45. van Zijl PCM, van Wezel RP, MacLean C, Bothner-By AA (1985) J. Phys. Chem. 89: 204
46. van Zijl PCM, Bastiaan EW, MacLean C, Bothner-By AA. To be published
47. Bastiaan EW, Wegman MAM, MacLean C (1987) Magn. Reson. Chem. 25: 817
48. Vander Hart DL, Flygare WH (1970) Mol. Phys. 18: 77
49. Battaglia MR, Ritchie GLD (1977) J. Chem. Soc. Far. Trans. II 73: 209
50. Vul'fson SG, Dianora OM (1984) Izv. Akad. Nauk. SSSR. Ser. Khim. 10: 2269 (Chem. Abstracts 102: 45459n)
51. Ritchie GLD, Cooper MK, Calvert RL, Dennis GR, Philips L, Vrbancich J (1983) J. Am. Chem. Soc. 105: 5215
52. Mulay LN, Boudreaux EA (1976) Theory and application of molecular diamagnetism. Wiley, New York
53. Lindsey JS, Mauzerall DC (1982) J. Am. Chem. Soc. 104: 4498
54. Lindsey JS, Mauzerall DC, Linschitz H (1983) J. Am. Chem. Soc. 105: 6528
55. Appleman BR, Dailey BP (1974) in: Waugh JS (ed). Academic, New York, pp 231–320 (Advances in magnetic resonance, vol 7)
56. Ditchfield R (1972) in: Buckingham AD, Allen G (eds). Butterworths, London, pp 91–157 (International review of science: physical chemistry)
57. Beans JW (1932) Rev. Mod. Phys. 4: 133
58. Flygare WH (1974) Chem. Rev. 74: 653
59. Lonsdale K, Krishnan KS (1936) Proc. R. Soc. Lond. A156: 597
60. Alms GR, Bauer DR, Brauman JI, Pecora R (1973) J. Chem. Phys. 59: 5304; 59: 5310
61. Patterson GD, Griffiths JE (1975) J. Chem. Phys. 63: 2406
62. Lewis D, Peters D (1975) Facts and theories of aromaticity. MacMillan, London
63. van Zijl PCM, Jenneskens LW, Bastiaan EW, MacLean C, de Wolf WH, Bickelhaupt F (1986) J. Am. Chem. Soc. 108: 1415
64. Haddon RC, Haddon VR, Jackman LM (1971) Top. Curr. Chem. 16: 103
65. Godfrey R (1978) J. Chem. Soc. Perkin Trans. II 10: 1019
66. Kidd KG, Kotowycz G, Schaefer T (1967) Can. J. Chem. 45: 2155
67. Schmalz TG, Norris CL, Flygare WH (1973) J. Chem. Phys. 95: 7961
68. Robert JM, Evilia RF (1985) J. Am. Chem. Soc. 107: 3733
69. Lamb DM, Vander Velde DG, Jonas J (1987) J. Magn. Reson. 73: 345

Behaviour of the NMR Relaxation Parameters at High Fields

D. Canet[1] **and J. B. Robert**[2]

[1] Laboratoire de Méthodologie RMN (UA CNRS 406-LESOC), Université de Nancy I, BP 239. 54506 Vandoeuvre-les-Nancy, Cedex France
[2] Service National des Champs Intenses CNRS, Avenue des Martyrs 38042 Grenoble Cedex France. Faculty Member of the Université J. Fourier Grenoble I

Table of Contents

NMR Basic Principles and Progress, Vol. 25

The algebraic expressions of the dynamic NMR parameters, namely, the spin–lattice relaxation time T_1, the spin–spin relaxation time T_2, the cross-relaxation terms σ and σ_ρ, and the NOE factor are derived. The study of their behaviour as a function of the applied magnetic field or resonance frequency is examined, with special emphasis on the high frequency region ($\nu > 800$ MHz). The different models which may be used to derive the spectral density are presented. Several plots of the variation of the NMR dynamic parameters for the most often encountered nuclei (^1H, ^{13}C, ^{15}N, ^{31}P) are shown. These plots correspond to different and realistic values of the correlation times, internuclear distances, and chemical shielding anisotropy. The variation of the NMR signal sensitivity as a function of the magnetic field is also discussed. As expected, with the exception of the phosphorus nucleus, the sensitivity increases with B_0. In some cases, the increase is greater than the expected factor $B_0^{3/2}$.

1 Introduction

The NMR parameters may be classified as static parameters and dynamic parameters. The static parameters, namely the indirect spin–spin coupling tensors **J** and the nuclear magnetic shielding tensors $\boldsymbol{\sigma}$, which reflect the electron behaviour in the molecules, may be considered as field independent in magnetic fields which have been used until now in high resolution NMR ($B_0 \leqslant 14T$; $\nu(^1H) \leqslant 600$ MHz). A field dependence of J and σ has been proposed only in very few cases [1], but nothing special is expected up to 23.5 T ($\nu(^1H) \sim 1$ GHz), the highest field strength which is considered in this chapter. In contrast to the static parameters, the dynamic ones (i.e. the spin–lattice relaxation time T_1, the spin–spin relaxation time T_2, the spin-locking relaxation time $T_{1\rho}$, and the Nuclear Overhauser Effect (NOE)), which depend on the molecular motion are field dependent.

When these dynamic NMR parameters are measured at different frequencies, information may be obtained about the molecular motions. However the magnetic resonances (NMR; EPR) are not the only experimental techniques which may provide information on molecular dynamics. There exist several other techniques which investigate both the individual and collective motions. Some of these techniques and their main features are summarized below [2]. They concern: (i) Coherent and incoherent neutron scattering; the coherent scattering reflects the collective atomic motions, the incoherent scattering reflects the individual rotational and translational motions. (ii) Absorption and scattering of electromagnetic waves. Dielectric absorption, which basically reflects the collective motions, is well adapted to the study of critical phenomena. Infrared absorption and Raman scattering are powerful methods for studying vibrational and rotational motions which may sometimes be difficult to separate. Rayleigh scattering of light is more suitable for the study of collective translational and vibrational motions. (iii) Other less commonly known techniques are used in some specific cases, such as resonant Raman scattering or the Mössbauer effect.

As compared to the above-mentioned methods, NMR spectroscopy has some advantages and particular features concerning the analysis of the molecular motions. In a given molecule using, if necessary, isotopic nuclear substitutions, the relaxation times or the NOE of several nuclei may be studied to gain molecular dynamics information. The labelling technique is interesting as it may allow the study of nuclei which occupy different sites with respect to the molecular symmetry elements. On the other hand, the nuclear relaxation samples different kinds of motion, depending on their nature. Thus the dipolar contribution to the relaxation time probes the reorientation of internuclear vectors, and the so-called CSA (Chemical Shift Anisotropy) mechanism samples the reorientation of a tensor principal axes with respect to the laboratory frame. Owing to the fact that NMR is a Zeeman spectroscopy, the spectra obtained at different magnetic field strengths permit sampling of the various Fourier components of the molecular motion in a wide frequency range, including the

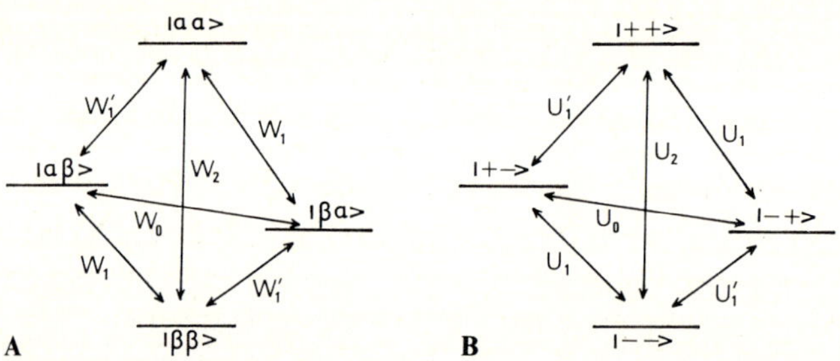

Fig. 1. Energy diagram for a two $\frac{1}{2}$ spin system. **A**) Transition probabilities between the eigenstates of longitudinal components of the spin operators. **B**) Transition probabilities between the eigenstates of the transverse components of the spin operators.

zero frequency which contributes to the spin–spin relaxation time T_2 and to the NOE. For $T_{1\rho}$, the frequency which contributes to the relaxation process is one or two times $\gamma B_1/2\pi$, where B_1 is the applied radio-frequency field in the rotating frame.

From a phenomenological point of view, for a set of mutually independent nuclei coupled to a thermal bath, the relaxation times T_1 and T_2 may be described by Bloch equations [3]. Denoting by M_0 the equilibrium magnetization along the direction of the static magnetic field direction B_0 (z direction) and by M_x, M_y, and M_z the instantaneous magnetization components along the x, y, z axes in the laboratory frame or in the rotating frame, one has:

$$\frac{dM_z}{dt} = \frac{M_0 - M_z}{T_1}$$

$$\frac{dM_x}{dt} = -\frac{M_x}{T_2} \quad \frac{dM_y}{dt} = -\frac{M_y}{T_2} \tag{1}$$

T_1 governs the magnetization evolution along B_0 and T_2 reflects the loss of spin coherence in a plane perpendicular to B_0. $T_{1\rho}$ refers to the magnetization evolution along the radio-frequency field B_1. Equation (1) leads to a monoexponential behaviour for M_z, M_x and M_y.

For a system of two coupled spins the evolution of the magnetization components is described by the Solomon equations [4]. The transition probabilities per unit time between the four levels of two 1/2 spin I and S are shown in Fig. 1. Note that the transitions considered are between the eigenstates of the magnetization operators $\hat{I}_z + \hat{S}_z$ (case A) and $\hat{I}_x + \hat{S}_x$ or $\hat{I}_y + \hat{S}_y$ (case B).

Considering the population evolution of the different levels in cases A and B respectively one has

$$\frac{d\langle \hat{I}_z \rangle}{dt} = -\frac{1}{T_{1I}}(\langle \hat{I}_z \rangle - I_0) - \sigma(\langle \hat{S}_z \rangle - S_o) \tag{2A}$$

$$\frac{d\langle \hat{S}_z \rangle}{dt} = -\frac{1}{T_{1S}}(\langle \hat{S}_z \rangle - S_o) - \sigma(\langle \hat{I}_z \rangle - I_o)$$

$$\frac{1}{T_{1I}} = W_o + 2W_1 + W_2; \quad \frac{1}{T_{1S}} = W_o + 2W'_1 + W_2; \quad \sigma = W_2 - W_o$$

$$\frac{d\langle \hat{I}_x \rangle}{dt} = -\frac{1}{T_{2I}} \langle \hat{I}_x \rangle - \sigma' \langle \hat{S}_x \rangle$$

$$\frac{d\langle \hat{S}_x \rangle}{dt} = -\frac{1}{T_{2S}} \langle \hat{S}_x \rangle - \sigma' \langle \hat{I}_x \rangle \tag{2B}$$

$$\frac{1}{T_{2I}} = (U_0 + 2U_1 + U_2) \quad \sigma' = (U_2 - U_0)$$

$$\frac{1}{T_{2S}} = (U_0 + 2U'_1 + U_2)$$

σ and σ' are named parallel and perpendicular cross relaxation terms respectively. The relaxation parameters are expressed here in terms of transition probabilities, and will be discussed in more detail in Sect. 2.

Contrary to the Bloch equations which show that the magnetization components follow a single exponential decay, the Solomon equations show that the decay for $\langle \hat{I}_z \rangle$ and $\langle \hat{S}_z \rangle$ is in general a linear combination of two exponentials. It must be pointed out that a single exponential decay is obtained if one deals with two identical nuclei, or if one of the two spins relaxes much faster than the other one, for example through an electron dipolar interaction. In case of more than two coupled spins, population evolution laws similar to the ones obtained from schemes A and B of Fig. 1 are derived. The transition probabilities can be obtained into a matrix W, which describes the population evolution collected into a vector (see Eq. 10).

In the following two sections a general derivation of the formulation of the relaxation mechanisms is presented along with a survey of the models which may be used to describe the molecular motions in the liquid phase. In the last two sections, the variation of sensitivity and of the dynamic NMR parameters are discussed, as a function of the magnetic field strength.

2 Formulation of Relaxation Parameters

This section is intended to provide the essential relaxation parameters, expressed with coherent and ready-to-use notations, in a way that permits their evolution with the applied static field B_0 and/or with the measurement frequency $\omega_0/2\pi$ to be easily followed. We shall first outline relaxation mechanisms (interactions) and formulate the relevant hamiltonian which constitutes the basis of the calculation of spectral densities. Any relaxation parameter can be expressed in terms of these quantities, which are the key for delineating the salient features of relaxation phenomena at high field.

2.1 Relaxation Mechanisms

The time dependent hamiltonian $\hat{H}_R(t)$ responsible for relaxation phenomena can, in any situation, be written in the form:

$$\hat{H}_R(t) = \sum_r \sum_{m=-2}^{+2} (\hat{A}_r^m)^+ F_r^m(t) - \overline{\hat{H}(t)} \tag{3}$$

(The symbol $^+$ means transposed complex conjugate). $\overline{\hat{H}(t)}$ is an ensemble average of the total hamiltonian and its inclusion in Eq. (3) forces the ensemble average of $\hat{H}_R(t)$ to be zero, which is an essential feature for deriving evolution equations. $\overline{\hat{H}(t)}$ concerns time dependent interactions whose ensemble average does not vanish and which consequently governs line positions and intensities in the NMR spectrum; these include in all cases shielding and indirect coupling tensors, and, for an anisotropic medium, dipolar and quadrupolar tensors. Subscript r denotes the interaction (mechanism) and implicitly defines the rank of the irreductible tensors $\hat{A}_r^m$ and F_r^m involved in (3). The rank of the considered tensors never exceeds two; m is the projection whose integer value may therefore lie between -2 and $+2$. One has from the general properties of irreductible tensors [5]: $(\hat{A}_r^m)^+ = (-1)^m \hat{A}_r^{-m}$ and $F_r^{-m} = (-1)^m (F_r^m)^*$. $\hat{A}_r^m$ are made of spin operators or linear combinations of operator products (see Table 1). They govern spin dynamics and especially the nature of recoveries after an initial perturbation which makes possible the evolution of nuclear magnetizations and therefore the measurement of dynamical parameters. The role played by the spin operators $\hat{A}_r^m$ appears in the general evolution equation [6]:

$$\frac{d\hat{\sigma}}{dt} = -i[\hat{H}_0, \hat{\sigma}] - \sum_{r,r'} \sum_{m=-2}^{2} [(\hat{A}_r^m)^+[\hat{A}_{r'}^m, \hat{\sigma} - \hat{\sigma}_{eq}]]\mathcal{J}^{rr'}(m\omega_0) \tag{4}$$

$\hat{\sigma}$ is the density operator and $\hat{H}_0$ the static hamiltonian generally identified with $\overline{\hat{H}(t)}$. The first term, on the right-hand side of the above equation simply represents nuclear precession. The second term accounts for relaxation processes. Since $\hat{\sigma}$ can be expressed in a spin operator basis, the double commutator involving the $\hat{A}_r^m$ operators acts just as a weighting factor of the spectral density $\mathcal{J}^{rr'}(m\omega_0)$. The spectral density only depends on the space functions F_r^m and $F_{r'}^m$, and on B_0 through the resonance frequency $\omega_0/2\pi$. In a general way, the evolution of any quantity $\langle \hat{M} \rangle$ can be deduced from $\hat{\sigma}$ through the trace relationship: $\langle \hat{M} \rangle = \mathrm{Tr}(\hat{\sigma}\hat{M})$ where $\hat{M}$ is the quantum mechanical operator associated with the considered quantity. Transforming (4) for $\mathrm{Tr}(\hat{\sigma}\hat{M})$ leads to:

$$\begin{aligned}\frac{d\langle \hat{M} \rangle}{dt} &= -i\mathrm{Tr}\{\hat{\sigma}[\hat{H}_0, \hat{M}]\} \\ &\quad - \sum_{r,r'} \sum_{m=-2}^{2} \mathrm{Tr}\{(\hat{\sigma} - \hat{\sigma}_{eq})[\hat{A}_{r'}^m, [(\hat{A}_r^m)^+, \hat{M}]]\}\mathcal{J}^{rr'}(m\omega_0)\end{aligned} \tag{5}$$

Evaluation of the commutators of the above equation yields, via the trace

Table 1. Spin operators and space functions (expressed in rad s^{-1}), in their irreducible form, for the relaxation mechanisms considered here. Spherical harmonies of rank two are used according to the following conventions: $Y_2^{\pm 2} = \sqrt{15/32\pi}\sin^2\theta\exp(\pm 2i\varphi)$, $Y_2^{\pm 1} = \mp\sqrt{15/32\pi}\sin 2\theta\exp(\pm i\varphi)$, $Y_2^0 = \sqrt{10/32\pi}(3\cos^2\theta - 1)$

Mechanism	$\hat{A}^{\pm 2}$	$\hat{A}^{\pm 1}$	$\hat{A}^0$	$F^{\pm 2}$	$F^{\pm 1}$	F^0
Random fields (1)		$\mp(1/\sqrt{2})\hat{I}_\pm$	$\hat{I}_z$		$\mp(\gamma/\sqrt{2})$ $(B_x \mp iB_y)$	γB_z
Chemical shift Anisotropy (general case) (2)		$\mp(\sqrt{3/8})\hat{I}_\pm$	$\hat{I}_z$		$\mp\sqrt{2/3}\gamma B_0$ $(\sigma_{xz} \pm i\sigma_{yz})$	$\gamma B_0(\sigma_{zz} - \sigma_0)$
Chemical shift Anisotropy (axial symmetry) (3)		$\mp(\sqrt{3/8})\hat{I}_\pm$	$\hat{I}_z$		$\sqrt{16\pi/45}\gamma B_0$ $\Delta\sigma Y_2^{\pm 1}(\theta,\varphi)$	$\sqrt{16\pi/45}\gamma B_0$ $\Delta\sigma Y_2^0(\theta,\varphi)$
Dipolar interaction (4)	$(\hat{I}_\pm^A\hat{I}_\pm^X)$	$\mp(\hat{I}_\pm^A\hat{I}_z^X + \hat{I}_z^A\hat{I}_\pm^X)/2$	$[4\hat{I}_z^A\hat{I}_z^X - (\hat{I}_+^A\hat{I}_-^X + \hat{I}_-^A\hat{I}_+^X)]/\sqrt{24}$	$K_D Y_2^{\pm 2}(\theta,\varphi)$	$K_D Y_2^{\pm 1}(\theta,\varphi)$	$K_D Y_2^0(\theta,\varphi)$
Quadrupolar interaction (general) (5)	$\hat{I}_\pm^2/2$	$\mp(\hat{I}_z\hat{I}_\pm + \hat{I}_\pm\hat{I}_z)/2$	$[4\hat{I}_z^2 - (\hat{I}_+\hat{I}_- + \hat{I}_-\hat{I}_+)]/\sqrt{24}$	$eQ'(V_{xx} - V_{yy} \pm 2iV_{xy})$	$\mp 2eQ'(V_{xz} \pm iV_{yz})$	$\sqrt{6}eQ'V_{zz}$
Quadrupolar interaction (axial symmetry) (6)	$\hat{I}_\pm^2/2$	$\mp(\hat{I}_z\hat{I}_\pm + \hat{I}_\pm\hat{I}_z)/2$	$[4\hat{I}_z^2 - (\hat{I}_+\hat{I}_- + \hat{I}_-\hat{I}_+)]/\sqrt{24}$	$K_Q Y_2^{\pm 2}(\theta,\varphi)$	$K_Q Y_2^{\pm 1}(\theta,\varphi)$	$K_Q Y_2^0(\theta,\varphi)$
Scalar relaxation of the second kind (7)		$\mp(1/\sqrt{2})\hat{I}_\pm$	$\hat{I}_z$		$\mp J_{AX}\hat{I}_\mp^X/\sqrt{2}$	$J_{AX}\hat{I}_z^X$

(1) γB in rad s^{-1}. B time dependent such as $\overline{B_x^2} = \overline{B_y^2} = \overline{B_z^2} = \overline{B^2}/3$, $\overline{B_xB_y} = \overline{B_xB_z} = \overline{B_yB_z} = 0$ and $\overline{B_x} = \overline{B_y} = \overline{B_z} = 0$.
(2) Elements of the shielding tensor (time dependent) are expressed in the laboratory frame. σ_0: one third of the shielding tensor trace. B_0: static magnetic field.
(3) x, y, z: molecular principal axis system of the shielding tensor; $|\sigma_{zz}| > |\sigma_{xx}| = |\sigma_{yy}|$; $\Delta\sigma = \sigma_{zz} - \sigma_{xx}$. θ and φ: polar angles (time dependent) of z with respect to the laboratory frame.
(4) $K_D = -(\mu_0/4\pi)\sqrt{24\pi/5}\hbar\gamma_A\gamma_X/r_{AX}^3$. θ and φ: polar angles of the direction joining the two interaction nuclei A and X, with respect to the laboratory frame.
(5) $Q' = Q/[4\hbar I(I+1)]$. eQ: nuclear quadrupole moment; I: nuclear spin number, V: electric field gradient tensor expressed in the laboratory frame.
(6) $K_Q = \sqrt{24\pi/5}eQV_{zz}/[4\hbar I(2I-1)]$; x, y, z: molecular principal axis system of the field gradient tensor; $|V_{zz}| > |V_{xx}| = |V_{yy}|$; θ and φ: polar angles (time dependent) of z with respect to the laboratory frame.
(7) J_{AX} in rads s^{-1}; $\hat{I}_X$ supposed to be time dependent through fast relaxation (A and X can be reversed)

including $\hat{\sigma}$, the quantities coupled to $\langle \hat{M} \rangle$ by relaxation. Also, $m\omega_0$ must be understood as ω_{ij}, the energy difference expressed in rad s^{-1}, between the $\hat{H}_0$ eigenstates $|i\rangle$ and $|j\rangle$ satisfying $\langle i|\hat{A}^m_{r'}|j\rangle \neq 0$.

Alternatively, Eq. (5) can be recast in a matrix form by means of the widely used Redfield equations [7]:

$$\frac{d}{dt}\sigma_{aa'}(t) = i(\omega_{a'} - \omega_a)\sigma_{aa'}(t) + \sum_{b,b'} R_{aa'bb'}\sigma_{bb'}(t) \tag{6}$$

where $\sigma_{aa'}$ is the (a, a') element of the matrix associated with the operator $(\hat{\sigma} - \hat{\sigma}_{eq})$. The density matrix is implicity expressed in the basis of $\hat{H}_0$ eigenvectors, so that $\omega_{a'}$ and ω_a represent the energy of eigenstates $|a'\rangle$ and $|a\rangle$. The most attractive feature of Redfield theory is the form of elements $R_{aa'bb'}$ which are expressed in a simple symmetric way:

$$R_{aa'bb'} = J_{aba'b'}(\omega_{a'} - \omega_{b'}) + J_{aba'b'}(\omega_a - \omega_b) - \delta_{a'b'}\sum_c J_{cbca}(\omega_c - \omega_b)$$
$$- \delta_{ab}\sum_c J_{ca'cb'}(\omega_c - \omega_{b'}) \tag{7}$$

with

$$J_{aba'b'}(\omega_{a'} - \omega_{b'}) = (-1)^m \sum_{r,r'} \mathscr{J}^{rr'}(\omega_{a'} - \omega_{b'})(\hat{A}_r^{-m})_{ab}(\hat{A}^m_{r'})_{b'a'} \tag{8}$$

In addition $R_{aa'bb'}$ is significantly different from zero if $|\omega_a - \omega_{a'}| \simeq |\omega_b - \omega_{b'}|$. An immediate consequence is the absence of coupling by relaxation between coherences ($|a\rangle \neq |a'\rangle$) and populations ($|a\rangle = |a'\rangle$). In other words, longitudinal magnetization does not couple into transverse magnetization. The evolution of a coherence is especially simple:

$$\frac{d}{dt}\sigma_{aa'} = [i(\omega_{a'} - \omega_a) + R_{aa'aa'}]\,\sigma_{aa'} \tag{9}$$

with

$$R_{aa'aa'} = 2\,J_{aaa'a'}(0) - \sum_c [J_{caca}(\omega_c - \omega_a) + J_{ca'ca'}(\omega_c - \omega_{a'})]$$

The term $i(\omega_{a'} - \omega_a)$ is associated with the precession frequency of the considered coherence, whereas $R_{aa'aa'}$ represents its specific relaxation rate and is related to its linewidth, provided that inhomogeneity effects of the static field can be neglected. It should be noticed that Eq. (9) predicts a monoexponential transverse relaxation. This is true as long as the considered coherence is well isolated. As soon as there exists a coherence $\sigma_{bb'}$ of frequency identical or close to that of $\sigma_{aa'}$, the term $R_{aa'bb'}$ couples both coherences.

Redfield equations are also appealing for longitudinal relaxation. Noticing that σ_{aa} can be identified with the population P_a of the eigenstate $|a\rangle$, or actually with its deviation from its equilibrium value, one has:

$$\frac{d}{dt}P_a = \sum_b W_{ab}P_b \tag{10}$$

where the familiar transition probabilities W_{ab} have been introduced:

$$W_{ab} = 2J_{abab}(\omega_a - \omega_b) \quad \text{for } a \neq b$$
$$W_{aa} = -\sum_{b \neq a} W_{ab} \tag{11}$$

In this chapter we shall essentially concentrate on the quantities $F_r^m(t)$ which concern space coordinates and which therefore contain information about molecular dynamics. The simple form of Eq. (3), where the ranks r of both irreductible tensors $\hat{A}_r^m$ and $F_r^m(t)$ are identical, stems from the general properties of irreductible tensors and is a consequence of the invariance of the considered interactions under any arbitrary rotation. The various mechanisms giving rise to spin relaxation are listed in Table 1; they all fulfill the above criteria with the exception of chemical shift anisotropy (CSA) [8], which indeed is not rotationally invariant since it involves the preferential direction of the static field B_0. Nevertheless, even in that case, a functional form similar to (3) can be devised: F_r^m remains an irreductible tensor of *rank two* whereas the $\hat{A}_r^m$ resemble (but are not identical) to irreductible tensor operators of *rank one.* m therefore goes from -1 to $+1$.

Some mechanisms do not appear in Table 1, for example intermolecular dipolar interactions as well as spin rotation, because they may be, to a good approximation, incorporated in randomly fluctuating fields. On the other hand, paramagnetic relaxation, which originates from interactions between the considered nuclear spin and an unpaired electron spin, may be accounted for by two mechanisms mentioned in Table 1: i) "scalar relaxation of the second kind" where J_{AX} is replaced by the hyperfine coupling, $\hat{I}^X$ operators being substituted by the electron spin operators $\hat{S}$ (contact term); ii) "random field" which includes the direct dipole–dipole interaction between nuclear and electron spins (pseudo-contact term).

Other exotic mechanisms [9], which may exist but whose experimental manifestations have not been observed up to now, will be disregarded. They include the antisymmetric part of the **J** and $\boldsymbol{\sigma}$ tensors, as well as the symmetric part of the **J** tensor, which anyway is unlikely to depend on B_0.

2.2 *Spectral Densities*

Second order perturbation theory leads to the conclusion that a given mechanism is active if two conditions are fulfilled: (i) The double commutator $[\hat{A}_{r'}^m, [\hat{A}_r^{m+}, \hat{M}]]$ of Eq. (5) must be non-zero, where $\hat{A}_{r'}^m$ and $\hat{A}_r^m$ are the operators corresponding to relaxation mechanisms, and where $\hat{M}$ is the magnetization of interest, taken here in a broad sense: longitudinal and transverse components are represented by $\hat{I}_z, \hat{I}_x$ and $\hat{I}_y$; $2\hat{I}_z^A\hat{I}_z^X$ is associated with a situation where one of the lines of the A doublet, in a two spin 1/2 AX system, has been inverted, etc... (ii) The correlation function of the space functions $F_r^m(t)$ must possess a non-zero component at frequencies related to the measurement frequency in order to induce transitions, which will consequently contribute to the time evolution of the relevant magnetizations.

This latter condition suggests the advisability of calculating the Fourier transform of correlation functions, which is termed the spectral density. Its general form, in an isotropic medium, is given by [10]:

$$\mathcal{J}^{rr'}(\omega) = \int_0^\infty \overline{F_r^0(t)F_{r'}^0(0)}\exp(-i\omega t)\,dt \tag{12}$$

where the bar denotes an ensemble average. Equation (12) deserves some comments:

(i) if r and r′ represent the same mechanism, the spectral density will be dubbed "autocorrelation";
(ii) if r and r′ refer to different mechanisms, they must be of *identical rank*. This situation corresponds to "cross-correlation" spectral densities [11, 12]. Usual examples include:
 —two distinct dipolar interactions (for instance, in a three spin system AXY: $\mathcal{J}^{d(AX,AY)}$).
 —one dipolar interaction and the CSA mechanism of one of the nuclei involved in the dipolar interaction (for instance $\mathcal{J}^{d(AX),CSA(A)}$ to be dealt with in the following).
(iii) for an isotropic medium, spectral densities are independent of the projection m. Therefore, provided that the double commutator $[\hat{A}_{r'}^{m+}, [\hat{A}_r^m, \hat{M}]]$ is non-zero, the spectral density given by Eq. (12) can be used regardless of the value of m.

Exploitation of Eq. (12) requires, in the general case, that F_r^0 and $F_{r'}^0$ be expressed in a molecular frame. The time dependence is then transferred to the angles defining the molecular frame with respect to the laboratory frame. This transformation is usually accomplished through the well-known Wigner matrices which apply to spherical irreductible tensors. Therefore, a motional model (see Sect. 3) is necessary for treating the correlation functions of Wigner matrices [13]. In all cases, the importance of a given mechanism is dictated by the magnitude of the "space functions" F_r^m. In some instances, the "space function" involves in fact spin operators, e.g. scalar relaxation of the second kind (see the mechanism of Table 1). Some simplifications occur in two particular situations: when dealing with an isotropic motion (that is, when the molecular entity can be considered as having a spherical shape) or when the tensor of the relevant mechanism is axially symmetric. We derive below the practical formulae corresponding to these two situations.

2.2.1 Isotropic Motion

We have first to express F_r^0 in that molecular frame which diagonalizes the tensor describing the considered interaction. For our purposes, we shall limit ourselves to tensors of rank 2, which correspond to the most common intramolecular mechanisms, and introduce the following notation:

—T_{r0} represents one third of the tensor trace.
—T_r is the largest element of the tensor, expressed in the (x, y, z) molecular frame where the tensor is diagonal; T_r corresponds to the z direction of that frame. $(T_r - T_{r0})$ can be deduced from the penultimate column of Table 1 by identifying the Z axis (of the laboratory frame) with the principal axis z of the molecular frame.
—η_r: the asymmetry parameter, $\eta_r = (T_{xx} - T_{yy})/T_r$ with $|T_r| \geqslant |T_{yy}| \geqslant |T_{xx}|$.

With these notations, and using Wigner matrices, F_r^0 can be expressed as:

$$F_r^0 = [(3\cos^2\beta - 1)/2](T_r - T_{r0}) + [(\sin^2\beta\cos 2\gamma)/2]\eta_r T_r \tag{13}$$

where β and γ are the Eulerian angles of the transformation from the laboratory frame (X, Y, Z) to the molecular frame (x, y, z) [5]. These angles are the only quantities that remain time dependent. Since β is the angle between Z and z, $\cos^2\beta$ can be, in a short-hand notation, identified with z^2 whereas $\sin^2\beta\cos 2\gamma$ can be identified with $(x^2 - y^2)$.

For convenience, we shall first consider an autocorrelation spectral density; according to (12), it can be written as $\mathscr{J}(F_r^0 \circ F_r^0)$; the symbol $\circ$ denotes a product of convolution. Likewise, we shall denote by $\mathscr{J}(z^2 \circ z^2)$ the Fourier transform of the correlation function $\overline{\cos^2\beta(t)\cos^2\beta(0)}$:

$$\mathscr{J}(z^2 \circ z^2) = \int_0^\infty \overline{\cos^2\beta(t)\cos^2\beta(0)}\exp(-i\omega t)\,dt \tag{14}$$

Since isotropic reorientation is assumed: $\mathscr{J}(x^2 \circ x^2) = \mathscr{J}(y^2 \circ y^2) = \mathscr{J}(z^2 \circ z^2)$. Other spectral densities involved in the evaluation of (12) by means of the expansion (13) are of the type $\mathscr{J}(z^2 \circ 1)$ and $\mathscr{J}(x^2 \circ y^2)$. The former is zero because it corresponds to the Fourier transform of a constant quantity, namely $\overline{\cos^2\beta}$. The latter can be evaluated by realizing that $x^2 + y^2 + z^2$ is a constant; as a consequence $\mathscr{J}[(x^2 + y^2 + z^2) \circ (x^2 + y^2 + z^2)] = 0$. Since all cross-terms of the type $\mathscr{J}(x^2 \circ y^2)$ are identical due to isotropic reorientation, expanding this equation yields: $\mathscr{J}(x^2 \circ y^2) = -\mathscr{J}(z^2 \circ z^2)/2$. Thus (12) can be expressed as a function of only $\mathscr{J}(z^2 \circ z^2)$, which itself, because of its dependency in $\cos\beta$ $(Y_2^0(\beta) = \sqrt{5/16\pi}(3\cos^2\beta - 1))$, can be rewritten by introducing to so-called *reduced spectral density* $\tilde{J}(\omega)$:

$$\tilde{J}(\omega) = 4\pi \int_{-\infty}^{+\infty} \overline{Y_2^0[\beta(t)]Y_2^0[\beta(0)]}\exp(-i\omega t)dt \tag{15}$$

Let us recall that, in the case of a monoexponential correlation function characterized by the correlation time τ_c, the real part of $\tilde{J}(\omega)$ reduces to $2\tau_c/(1 + \omega^2\tau_c^2)$. Its imaginary part is usually neglected, regardless of the nature of the correlation function. Now, inserting the reduced spectral density in the expression of an autocorrelation spectral density leads to:

$$\mathcal{J}^{r}(\omega) = (1/10)[(T_r - T_{r0})^2 + \eta_r^2 T_r^2/3]\tilde{J}(\omega) \tag{16}$$

The above development can be transposed to the case of a cross-correlation spectral density:

$$\mathcal{J}^{rr'}(\omega) = (1/10)[(T_r - T_{r0})(T_{r'} - T_{r'0}) + \eta_r \eta_{r'} T_r T_{r'}/3]\tilde{J}^{rr'}(\omega) \tag{17}$$

with the cross-correlation reduced spectral density:

$$\tilde{J}^{rr'}(\omega) = 4\pi \int_{-\infty}^{+\infty} \overline{Y_2^0[\beta_r(t)] Y_2^0[\beta_{r'}(0)]} \exp(-i\omega t)\, dt \tag{18}$$

which can be related to the conventional reduced spectral density by:

$$\tilde{J}^{rr'}(\omega) = [(3\cos^2\theta_{rr'} - 1)/2]\tilde{J}(\omega) \tag{18a}$$

$\theta_{rr'}$ is the angle between the two principal directions, z and z', of the two tensors r and r'. It must be emphasized that (18a) is only valid for an isotropic motion. It is easily deduced by expressing one of the spherical harmonics in (18), say $Y_2^0[\beta_{r'}(0)]$, in the principal axis system of the other tensor, denoted here by r.

2.2.2 *Axially Symmetric Tensor*

Relying on the transformation (13) with $\eta_r = 0$, and considering reduced spectral densities whose spherical harmonics Y_2^0 involve the angle between the principal symmetry axes of the considered tensor and the Z axis of the laboratory frame, we obtain:

$$\mathcal{J}^{r}(\omega) = (1/10)(T_r - T_{r0})^2 \tilde{J}(\omega) \tag{19}$$

$$\mathcal{J}^{rr'}(\omega) = (1/10)(T_r - T_{r0})(T_{r'} - T_{r'0})\tilde{J}^{rr'}(\omega) \tag{20}$$

Since the Eulerian angle γ disappears in formula (13), the assumption of an isotropic motion is no longer necessary and the spectral densities probe the reorientation of a given molecular vector called, in the following the relaxation vector. It is in the most frequently encountered cases: i) the internuclear vector for dipolar relaxation; ii) the symmetry axis of the shielding tensor for the CSA mechanism; iii) the symmetry axis of the field gradient tensor for quadrupolar relaxation.

2.3 *Relaxation Parameters*

We shall consider in this section the additive contribution of each mechanism to the longitudinal $R_1 = T_1^{-1}$ or the transverse $R_2 = T_2^{-1}$ relaxation rates when they are unambiguously defined, i.e. whenever evolutions are monoexponential. If this is not the case, the set of simultaneous differential equations leading to a multiexponential evolution will be set up.

2.3.1 *Relaxation by Random Fields*

To a first approximation, the considered interactions can be accounted for by randomly varying magnetic fields (see Table 1). Applying Eq. (5) with $\hat{M} = \hat{I}_z$ or $\hat{I}_x$, $\hat{I}_y$ and calculating the relevant commutators, we arrive at Bloch equations (monoexponential evolution):

$$d\langle \hat{I}_z \rangle/dt = -R_1^{rf}(\langle \hat{I}_z \rangle - I_{eq}) \tag{21}$$

$$d\langle \hat{I}_{x,y} \rangle/dt = -R_2^{rf}\langle \hat{I}_{x,y} \rangle \tag{22}$$

Precession terms have been omitted in (22). R_1 and R_2 can be expressed as a function of spectral densities defined by Eq. (12) by using the space functions F given in Table 1:

$$R_1^{rf} = 2\mathscr{J}^{rf}(\omega_0) \tag{23}$$

$$R_2^{rf} = \mathscr{J}^{rf}(0) + \mathscr{J}^{rf}(\omega_0) \tag{24}$$

where $\omega_0/2\pi$ is the resonance frequency of the considered nucleus. The static field acts through a possible frequency dependence of $\mathscr{J}^{rf}(\omega_0)$.

2.3.2 *Relaxation by Chemical Shift Anisotropy (CSA)*

Again, Bloch equations prevail since calculations of commutators in Eq. (5) with $\hat{M} = \hat{I}_z$ lead only to $\hat{I}_z$; an homologous behaviour is found for $\hat{I}_x$ and $\hat{I}_y$. These calculations are carried out with the spin operators of Table 1 and lead to:

$$R_1^{CSA} = (3/2)\mathscr{J}^{CSA}(\omega_0) \tag{25}$$

$$R_2^{CSA} = \mathscr{J}^{CSA}(0) + (3/4)\mathscr{J}^{CSA}(\omega_0) \tag{26}$$

Since we are dealing with an intramolecular interaction based on spherical harmonics of rank 2 (see Table 1), reduced spectral densities can be introduced. This yields for the case of an isotropic motion:

$$R_1^{CSA} = (1/15)(\gamma B_0)^2(\Delta\sigma)^2(1 + \eta_{CSA}^2/3)\tilde{J}(\omega_0) \tag{27}$$

$$R_2^{CSA} = (1/15)(\gamma B_0)^2(\Delta\sigma)^2(1 + \eta_{CSA}^2/3)[2\tilde{J}(0)/3 + \tilde{J}(\omega_0)/2] \tag{28}$$

with $\Delta\sigma = \sigma_{zz} - (\sigma_{xx} + \sigma_{yy})/2$ and $\eta_{CSA} = 3(\sigma_{xx} - \sigma_{yy})/(2\Delta\sigma)$; x, y and z are the principal *molecular axes* of the shielding tensor ($|\sigma_{zz}| \geqslant |\sigma_{yy}| \geqslant |\sigma_{xx}|$). η_{CSA}, as it stands, differs from the usual definition ($\eta = (\sigma_{xx} - \sigma_{yy})/\sigma_{zz}$) because one is dealing with a non-zero trace tensor [14]. Similar formulas hold for an axially symmetric tensor ($\eta = 0$) without the restriction of an isotropic motion for the evaluation of $\tilde{J}$. It must be emphasized that the dependence with respect to B_0 arises not only from $\tilde{J}(\omega_0)$ (which can be evaluated according to the nature of rotational motions) but also from the factor $(\gamma B_0)^2$. On the other hand, R_2^{CSA} never equals R_1^{CSA} (and is always larger): even in the extreme narrowing case ($\tilde{J}(\omega) = 2\tau_c$ for any ω), $R_2^{CSA}/R_1^{CSA} = 7/6$ (see Sect. 5). In this discussion, the contribution of the antisymmetric components of the $\boldsymbol{\sigma}$ tensor has been omitted.

2.3.3 Dipolar Interaction between Two Nuclei of Spin 1/2

The two interacting nuclei are denoted A and X; they are supposed to be weakly coupled (i.e. their indirect coupling constant J is small with respect to their difference in resonance frequencies). Application of Eq. (5) to $\hat{I}_z^A$ and $\hat{I}_z^X$ yields a set of two simultaneous differential equations, known as Solomon equations (see Eq. (2)) which of course entail a biexponential evolution for both $\langle \hat{I}_z^A \rangle$ and $\langle \hat{I}_z^X \rangle$:

$$d\langle \hat{I}_z^A \rangle/dt = -R_1^A(\langle \hat{I}_z^A \rangle - I_{eq}^A) - \sigma(\langle \hat{I}_z^X \rangle - I_{eq}^X)$$
$$d\langle \hat{I}_z^X \rangle/dt = -R_1^X(\langle \hat{I}_z^X \rangle - I_{eq}^X) - \sigma(\langle \hat{I}_z^A \rangle - I_{eq}^A) \quad (29)$$

R_1^A and R_1^X are the specific relaxation rates of nuclei A and X. Disregarding in this section the CSA mechanism, we shall consider them as arising from random fields and from dipolar interactions. The random field contribution, as defined by Eq. (23), can simply be added to the dipolar relaxation rate. The latter can be directly expressed in terms of reduced spectral densities since the dipolar tensor is, by nature, axially symmetric:

$$(R_1^A)^d = (1/20)(\mu_0/4\pi)^2(\gamma_A\gamma_X\hbar/r_{AX}^3)^2[6\tilde{J}(\omega_A+\omega_X) + 3\tilde{J}(\omega_A) + \tilde{J}(\omega_A-\omega_X)] \quad (30)$$

r_{AX} is the internuclear distance; all constants have their usual meaning: μ_0 is the magnetic permeability of vacuum, γ's are gyromagnetic ratios (in rad.s^{-1}.T^{-1}) and $\hbar$ the Planck constant divided by 2π. A similar expression holds for $(R_1^X)^d$. Whenever a dipolar interaction occurs with a nucleus X of arbitrary spin I_X, formula (30) transforms into:

$$(R_1^A)^d = (1/15)I_X(I_X+1)(\mu_0/4\pi)^2(\gamma_A\gamma_X\hbar/r_{AX}^3)^2 \times [6\tilde{J}(\omega_A+\omega_X) + 3\tilde{J}(\omega_A) + \tilde{J}(\omega_A-\omega_X)] \quad (30\,\text{bis})$$

σ is the so-called cross-relaxation term. It arises solely from dipolar interactions:

$$\sigma = (1/20)(\mu_0/4\pi)^2(\gamma_A\gamma_X\hbar/r_{AX}^3)^2[6\tilde{J}(\omega_A+\omega_X) - \tilde{J}(\omega_A-\omega_X)] \quad (31)$$

Whenever the two nuclei A and X become identical, or rather isochronous (like spins), their dipolar relaxation rate is obtained by adding $(R_1^A)^d$ and σ. This yields with $\omega_0 = \omega_A = \omega_X$:

$$(R_1^{like})^d = (1/20)(\mu_0/4\pi)^2(\gamma^2\hbar/r^3)^2[12\tilde{J}(2\omega_0) + 3\tilde{J}(\omega_0)] \quad (32)$$

In the extreme narrowing regime, Eq. (32) exhibits the well-known 3/2 effects [Ref. 6 p. 297], by comparison with the dipolar relaxation rate pertaining to a single nucleus (Eq. (30)). Note that an apparent relaxation rate, identical to the one of Eq. (32), also occurs for an homonuclear spin system of *unlike spins*, provided that the system undergoes non-selective perturbations (rf pulses).

σ gives rise to the *NOE Spectroscopy* [15]. In its simplest form it consists in saturating one of the nuclei, say A. A new "equilibrium" value, I_{st}^X is then reached by X. It is easily evaluated from Eq. (29) by setting $d/dt\langle \hat{I}_z^X \rangle$ and $\langle \hat{I}_z^A \rangle$

to zero:

$$I^X_{st} = I^X_{eq}[1 + (\gamma_A/\gamma_X)(\sigma/R^X_1)]$$

which leads to the widely used NOE factor:

$$\eta = (I^X_{st} - I^X_{eq})/I^X_{eq} = (\gamma_A/\gamma_X)(\sigma/R^X_1) \tag{33}$$

For both $(R_1)^d$ and σ, the dependence with respect to B_0 originates only from rotational motion through $\tilde{J}(\omega)$. It can be mentioned that this is still true for the so-called dipolar–dipolar *cross-correlation* spectral densities which occur for spin systems comprising more than two spins and which involve two distinct dipolar interactions sharing one identical nucleus (see Eqs. (18) and (20)).

Transverse relaxation rates are more easily obtained by means of Redfield equations. Let us suppose that, in addition to their dipolar interaction, whose ensemble average is zero, nuclei A and X are J-coupled so that A and X spectra appear in the form of doublets. Applying Eqs. (8) and (9) from Redfield theory, we obtain the random field and dipolar contributions to the relaxation rates. The following labeling (for an AX spin system) will be used throughout; eigenstates: $|1\rangle = \alpha(A)\alpha(X)$, $|2\rangle = \alpha(A)\beta(X)$, $|3\rangle = \beta(A)\alpha(X)$, $|4\rangle = \beta(A)\beta(X)$ whereas lines A_1, A_2, X_1 and X_2 will correspond to transitions 1–3, 2–4, 1–2 and 3–4, respectively. Recall that the linewidth of transition 1–2 is equal to R_{12}/π, where R_{12} stands for $-R_{1212}$ used in the Redfield equations:

$$R^d_{12} = R^d_{13} = R^d_{24} = R^d_{34} = (1/20)(\mu_0/4\pi)^2(\gamma_A\gamma_X\hbar/r^3_{AX})^2[2\tilde{J}(0) + \tilde{J}(\omega_A - \omega_X)/2 + 3\tilde{J}(\omega_A + \omega_X) + 3\tilde{J}(\omega_A)/2 + 3\tilde{J}(\omega_X)/2] \tag{34}$$

$$R^{rf}_{12} = R^{rf}_{34} = \mathscr{J}^{rf(X)}(0) + \mathscr{J}^{rf(X)}(\omega_X) + \mathscr{J}^{rf(A)}(\omega_A) \tag{35}$$

$$R^{rf}_{13} = R^{rf}_{24} = \mathscr{J}^{rf(A)}(0) + \mathscr{J}^{rf(A)}(\omega_A) + \mathscr{J}^{rf(X)}(\omega_X) \tag{35 bis}$$

In view of these expressions, it can be noticed that a factor of broadening acting specifically on spin A (e.g. a paramagnetic species in the vicinity of that spin) contributes indirectly to the linewidth of X transitions *via* $\mathscr{J}^{rf(A)}(\omega_A)$ which is also present in Eq. (35). This feature has been scarcely recognized [16]. Furthermore, Redfield theory implies a coupling term between transitions 1–2 and 3–4, as well as between transitions 1–3 and 2–4:

$$R^d_{1234} = (-1/20)(\mu_0/4\pi)^2(\gamma_A\gamma_X\hbar/r^3_{AX})^2(3\tilde{J}(\omega_A)/2) \tag{36}$$

$$R^d_{1324} = (-1/20)(\mu_0/4\pi)^2(\gamma_A\gamma_X\hbar/r^3_{AX})^2(3\tilde{J}(\omega_X)/2) \tag{36a}$$

$$R^{rf}_{1234} = \mathscr{J}^{rf(A)}(\omega_A) \tag{37}$$

$$R^{rf}_{1324} = \mathscr{J}^{rf(X)}(\omega_X) \tag{37a}$$

These coupling terms do not influence linewidths as long as the two lines in each doublet are well separated. However, if the J coupling becomes vanishingly small, the two lines in each doublet collapse into a single line whose linewidth will be deduced from the time evolution of the sum of two off-diagonal elements of the density matrix. For instance, one has for the X line:

$$d(\sigma_{12}+\sigma_{34})/dt = -R_{12}(\sigma_{12}+\sigma_{34}) + R_{1234}(\sigma_{12}+\sigma_{34})$$

which yields for the effective transverse rate:

$$R_2^X = (1/20)(\mu_0/4\pi)^2(\gamma_A\gamma_X\hbar/r_{AX}^3)^2[2\tilde{J}(0) + \tilde{J}(\omega_A-\omega_X)/2 + 3\tilde{J}(\omega_A+\omega_X) + 3\tilde{J}(\omega_X)/2 + 3\tilde{J}(\omega_A)] + \mathscr{J}^{rf(X)}(0) + \mathscr{J}^{rf(X)}(\omega_x) \quad (38)$$

in agreement with Abragam's result [Ref. 6 p. 296]. An homologous expression holds for R_2^A by interchanging labels A and X. Finally, the existence of a *transverse cross-relaxation* term defined in the introduction must be considered. Its expression, for two *isotopes of identical nature*, can be deduced from the calculations presented in Sect. 2.3.5:

$$\sigma_t = (1/20)(\mu_0/4\pi)^2(\gamma^2\hbar/r_{AX}^3)^2[2\tilde{J}(0) + 3\tilde{J}(\omega_0)]$$

σ_t is added to (38) in order to obtain the transverse relaxation rate pertaining to like spins:

$$(R_2^{like})^d = (1/20)(\mu_0/4\pi)^2(\gamma^2\hbar/r_{AX}^3)^2[9\tilde{J}(0)/2 + 15\tilde{J}(\omega_0)/2 + 3\tilde{J}(2\omega_0)] \quad (39)$$

The modulation of the dipolar interaction may also originate from an intermolecular translational motion which makes the internuclear distance time dependent. In that case the relevant relaxation parameters can be rewritten acccording to the general expression of spectral densities by droping $(r_{AX})^{-6}$ in Eqs. (30)–(39) and replacing $\tilde{J}(\omega)$ by

$$\tilde{J}_T(\omega) = 4\pi\int_{-\infty}^{+\infty} \overline{[Y_2^0(t)/r_{AX}^3(t)][Y_2^0(0)/r_{AX}^3(0)]}\exp(-i\omega t)dt$$

2.3.4 *Dipolar Interaction and Chemical Shift Anisotropy in a System of Two Spin 1/2 Nuclei*

Due to the structure of the space functions pertaining to dipolar interactions and CSA, which both involve irreductible tensors of rank 2, a non zero cross-correlation spectral density between these two mechanisms may exist. This cross-correlation spectral density is often referred to as an *interference* term. It turns out that, under these circumstances, Solomon equations do not fully account for the evolution of longitudinal magnetizations [17], [18]. The CSA–dipolar interference term couples $\langle\hat{I}_z^A\rangle$ and $\langle\hat{I}_z^X\rangle$ to a third quantity $2\langle\hat{I}_z^A\hat{I}_z^X\rangle$. This quantity, which arises from the double commutator in Eq. (5), will be seen to be associated with the difference in line intensities within each doublet. This additional term leads to generalized Solomon equations:

$$\begin{aligned} d\langle\hat{I}_z^A\rangle/dt &= -R_1^A(\langle\hat{I}_z^A\rangle - I_{eq}^A) - \sigma(\langle\hat{I}_z^X\rangle - I_{eq}^X) - \sigma'_A(2\langle\hat{I}_z^A\hat{I}_z^X\rangle) \\ d\langle\hat{I}_z^A\rangle/dt &= -R_1^X(\langle\hat{I}_z^X\rangle - I_{eq}^X) - \sigma(\langle\hat{I}_z^A\rangle - I_{eq}^A) - \sigma'_X(2\langle\hat{I}_z^A\hat{I}_z^X\rangle) \quad (40) \\ d(2\langle\hat{I}_z^A\hat{I}_z^X\rangle)/dt &= -R'_1(2\langle\hat{I}_z^A\hat{I}_z^X\rangle) - \sigma'_A(\langle\hat{I}_z^A\rangle - I_{eq}^A) - \sigma'_X(\langle\hat{I}_z^X\rangle - I_{eq}^X) \end{aligned}$$

R_1^A as well as R_1^X are the sum of dipolar, random field and CSA contributions as given by formulas (30), (23) and (27), respectively. σ has the same meaning as in classical Solomon equations (formula (31)), whereas σ'_A and σ'_X arise from the CSA–dipolar interference term:

$$\sigma'_A = -\sqrt{3}\mathscr{J}^{d(AX),CSA(A)}(\omega_A) \tag{41}$$

$$\sigma'_X = -\sqrt{3}\mathscr{J}^{d(AX),CSA(X)}(\omega_X) \tag{41a}$$

R'_1 represents the specific relaxation rate of the quantity $2\langle \hat{I}_z^A \hat{I}_z^X \rangle$:

$$R'_1 = 2\mathscr{J}^{rf(A)}(\omega_A) + (3/2)\mathscr{J}^{CSA(A)}(\omega_A) + 2\mathscr{J}^{rf(X)}(\omega_X) + (3/2)\mathscr{J}^{CSA(X)}(\omega_X) + (3/20)(\mu_0/4\pi)^2(\gamma_A\gamma_X\hbar/r_{AX}^3)^2[\tilde{J}(\omega_A) + \tilde{J}(\omega_X)] \tag{42}$$

$\mathscr{J}^{CSA(A)}(\omega_A)$ or $\mathscr{J}^{CSA(X)}(\omega_X)$ can be expressed in terms of reduced spectral densities as in Eq. (27) in the case of an isotropic motion or of an axially symmetric tensor. This can also be done for the interference term:

$$\mathscr{J}^{d(AX),CSA(A)}(\omega_A) = -\sqrt{2/75}(\mu_0/4\pi)(\gamma_A\gamma_X\hbar/r_{AX}^3)(\gamma_A B_0)(\Delta\sigma_A)\tilde{J}^{d(AX),CSA(A)}(\omega_A) \tag{43}$$

In the case of an isotropic motion, the cross-correlation reduced spectral density appearing in Eq. (43) can be expressed as a function of the conventional reduced spectral density, and by virtue of equation (18a):

$$\tilde{J}^{d(AX),CSA(A)}(\omega_A) = [(3\cos^2\theta_{AX,A} - 1)/2]\tilde{J}(\omega_A) \tag{44}$$

where $\theta_{AX,A}$ is the angle between the vector joining the two nuclei A and X, and the principal direction of the shielding tensor.

The question is now to understand how this new feature, i.e. the dipolar-CSA interference term, manifests itself in experiments aimed at measuring longitudinal relaxation. First of all, the evolution of each total longitudinal magnetization will depend, in the general case, on three exponentials, since Eq. (40) involve three simultaneous differential equations. It must however be borne in mind that this threefold exponentiality rests on a non-negligible value of the chemical shift anisotropy of the considered nucleus. For instance, if the CSA of nucleus A is small, σ'_A will be negligible and coupling with $2\langle \hat{I}_z^A \hat{I}_z^X \rangle$ disappears; evolution of A magnetization remains biexponential. This means that it is the nucleus possessing the largest CSA which entails a threefold exponentiality; this feature depends directly on B_0, as shown by Eq. (43), in addition to the field dependence of the reduced spectral density due to possible non-extreme narrowing conditions. In order to illustrate this feature, we notice from Eq. (43) that the interference term evolves with frequency (or applied field) as $2\omega_A\tau_c/(1 + \omega_A^2\tau_c^2)$ provided we choose the simplest form of the reduced spectral density (see Sect. 3). The quantity $2\omega\tau_c/(1 + \omega^2\tau_c^2)$ is plotted on Fig. 2 for $\tau_c = 10^{-9}$ s as a function of the measurement frequency. It can be seen that the CSA–dipolar interference term exhibits a broad maximum around 150 MHz. For other correlation times one has only to move the whole plot with respect to the frequency scale, as indicated in the figure.

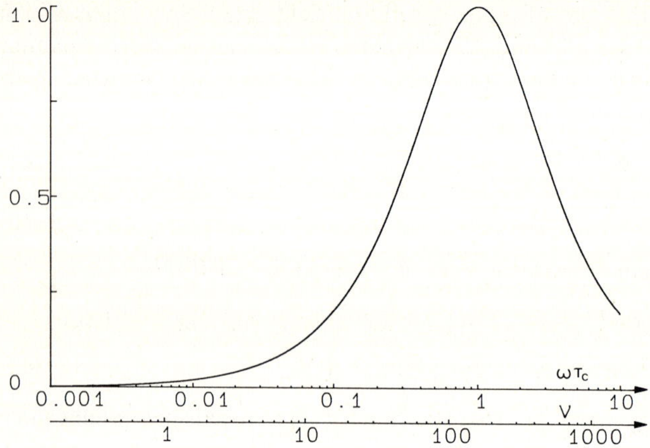

Fig. 2. Plot of the quantity $2\omega\tau_c/(1+\omega^2\tau_c^2)$, involved in the CSA–dipolar interference terms, as a function of:

1) $\omega\tau_c$ upper horizontal axis;
2) $\nu(\mathrm{MH_Z}) = (\omega/2\pi)$ for $\tau_c = 10^{-9}$ s, lower horizontal axis.

For using this representation for other τ_c values ($\tau_c = n \times 10^{-9}$ s), shift the whole curve by n frequency decades to the right if n is negative, and to the left is n positive

The other problem concerns the doublet structure for each spin A and X. It turns out that, under appropriate circumstances, the two lines within a given doublet evolve in a different way because of the interference term [19, 20]. In order to properly understand this effect, we must specify the nature of the observing pulse and the way in which it transforms the quantity $2\langle \hat{I}_z^A \hat{I}_z^X \rangle$. It is clear that a $(90°)_x$ non-selective pulse (acting in the same manner on A and X) will transform $\langle \hat{I}_z^A \rangle$ into $\langle \hat{I}_y^A \rangle$ and $\langle \hat{I}_z^X \rangle$ into $\langle \hat{I}_y^X \rangle$; in the same time $2\langle \hat{I}_z^A \hat{I}_z^X \rangle$ will be converted into $2\langle \hat{I}_y^A \hat{I}_y^X \rangle$, which is known to be associated with non-observable (zero and double quantum) coherences. Consequently, transverse magnetizations will exactly reflect the state of the corresponding longitudinal magnetizations at the time of the detection. Moreover, each doublet remains symmetrical. Consider now a semi-selective observing pulse acting only, say, on nucleus A. A $(90°)_x^A$ converts $2\langle \hat{I}_z^A \hat{I}_z^X \rangle$ into $2\langle \hat{I}_y^A \hat{I}_z^X \rangle$. The latter quantity is associated with the two lines A_1 and A_2 antiphase along the y axis of the rotating frame (A_1 pointing toward +y). Owing to the fact that $\langle \hat{I}_z^A \rangle$ has been converted into $\langle \hat{I}_y^A \rangle$, which is represented by the two lines A_1 and A_2 along +y, a differential evolution of lines A_1 and A_2 is expected. A similar phenomenon is also expected after the application of a non-selective observing pulse, whose flip angle differs from 90° [21, 22], since it will convert $2\langle \hat{I}_z^A \hat{I}_z^X \rangle$ partly into $2\langle \hat{I}_y^A \hat{I}_z^X \rangle$. This feature can be illustrated by looking at the initial behavior of lines A_1 and A_2. Suppose that, at the onset, A has been semi-selectively inverted; $2\langle \hat{I}_z^A \hat{I}_z^X \rangle$ is zero at thermal equilibrium and remains zero immediately after this inverting pulse. Consequently, one has:

$$(d\langle \hat{I}_z^A\rangle/dt)_{t\to 0} = R_1^A(2I_{eq}^A)$$

$$(d\langle 2\hat{I}_z^A\hat{I}_z^X\rangle/dt)_{t\to 0} = \sigma'_A(2I_{eq}^A)$$

which yields for the initial slopes s_0 of the two following quantities, associated with the intensities of lines A_1 and A_2:

$$s_0(A_1/2A_1^{eq}) = R_1^A + \sigma'_A \quad (45)$$

$$s_0(A_2/2A_2^{eq}) = R_1^A - \sigma'_A \quad (45a)$$

It is apparent that the CSA–dipolar interference term can be directly deduced from this differential behavior. If σ'_A is positive (its sign is determined by the signs of $\Delta\sigma_A$ and of $\tilde{j}^{d(AX),CSA(A)}(\omega_A)$; see Eqs. (43) and (44)), line A_1 relaxes faster than A_2. The relative position of the two lines further depends upon the sign of the indirect coupling constant J_{AX}. Therefore, a proper interpretation of a possible differential behavior of the two lines in a doublet requires knowledge of both signs of the coupling constant and of the interference term.

Related effects occur for linewidths. Applying Redfield theory allows to calculate the following contributions which must be added to R_{12}, R_{13}, R_{24}, R_{34}, R_{1234} and R_{1324} defined by Eqs. (34)–(37a):

$$R_{12}^{CSA} = \mathcal{J}^{CSA(X)}(0) + (3/4)\mathcal{J}^{CSA(X)}(\omega_X) + (3/4)\mathcal{J}^{CSA(A)}(\omega_A) + \sqrt{2/3}\,\mathcal{J}^{d(AX),CSA(X)}(0) + \sqrt{3/8}\,\mathcal{J}^{d(AX),CSA(X)}(\omega_X) \quad (46)$$

$$R_{34}^{CSA} = \mathcal{J}^{CSA(X)}(0) + (3/4)\mathcal{J}^{CSA(X)}(\omega_X) + (3/4)\mathcal{J}^{CSA(A)}(\omega_A) - \sqrt{2/3}\,\mathcal{J}^{d(AX),CSA(X)}(0) - \sqrt{3/8}\,\mathcal{J}^{d(AX),CSA(X)}(\omega_X) \quad (47)$$

$$R_{13}^{CSA} = \mathcal{J}^{CSA(A)}(0) + (3/4)\mathcal{J}^{CSA(A)}(\omega_A) + (3/4)\mathcal{J}^{CSA(X)}(\omega_X) + \sqrt{2/3}\,\mathcal{J}^{d(AX),CSA(A)}(0) + \sqrt{3/8}\,\mathcal{J}^{d(AX),CSA(A)}(\omega_A) \quad (46a)$$

$$R_{24}^{CSA} = \mathcal{J}^{CSA(A)}(0) + (3/4)\mathcal{J}^{CSA(A)}(\omega_A) + (3/4)\mathcal{J}^{CSA(X)}(\omega_X) - \sqrt{2/3}\,\mathcal{J}^{d(AX),CSA(A)}(0) - \sqrt{3/8}\,\mathcal{J}^{d(AX),CSA(A)}(\omega_A) \quad (47a)$$

$$R_{1234}^{CSA} = (3/4)\mathcal{J}^{CSA(A)}(\omega_A) - \sqrt{3/32}\,\mathcal{J}^{d(AX),CSA(A)}(\omega_A) \quad (48)$$

$$R_{1324}^{CSA} = (3/4)\mathcal{J}^{CSA(X)}(\omega_X) - \sqrt{3/32}\,\mathcal{J}^{d(AX),CSA(X)}(\omega_X) \quad (48a)$$

Again, differential broadening due to the dipolar–CSA interference term can occur within each doublet (see for instance Eqs. (46) and (47)). On the other hand, in a manner anologous to random field effects, *widths of A transitions depend on Chemical Shift Anisotropy at X*. This feature, of course, disappears when the coupling constant vanishes, i.e. when the two lines of the doublet collapse. In such a case, it becomes possible to define a transverse relaxation rate as was done before by using R_{1234} and R_{1324}. One has to add the following quantity to the transverse relaxation rate defined previously by Eq. (38):

$$(R_2^X)^{CSA} = \mathcal{J}^{CSA(X)}(0) + (3/4)\mathcal{J}^{CSA(X)}(\omega_X) + \sqrt{2/3}\,\mathcal{J}^{d(AX),CSA(X)}(0) + \sqrt{3/8}\,\mathcal{J}^{d(AX),CSA(X)}(\omega_X) + \sqrt{3/32}\,\mathcal{J}^{d(AX),CSA(A)}(\omega_A) \quad (49)$$

Note that dipolar–CSA interference terms are still present in this transverse relaxation rate. This is not surprising since the two nuclei A and X, even in the absence of J coupling, still "communicate" through dipolar interaction.

2.3.5 *Relaxation in the Rotating Frame*

The experimental conditions for carrying out relaxation measurements in the rotating frame require, during the evolution period, a radiofrequency field polarized along, say, the x direction, which is supposed to lock the nuclear magnetization along the same direction. We shall further assume that the strength of this rf field, measured by $\omega_1 = \gamma B_1$, is much larger than $|\omega - \omega_A|$ and/or $|\omega - \omega_X|$ where $\omega/2\pi$ is the frequency of the spin-lock field, and $\omega_A/2\pi$ and $\omega_X/2\pi$ the resonance frequencies of A and X, respectively. In the case of an homonuclear spin system, this means that $\omega_1 \gg |\omega_A - \omega_X|$. Conversely, in the case of a heteronuclear two spin system, where a spin-lock field with the same ω_1 is supposed to be applied to both nuclei, one has $\omega_1 \gg |\omega'_A - \omega_A|$ and $\omega_1 \gg |\omega'_X - \omega_X|$, where ω'_A and ω'_X stand for the actual frequencies of the spin lock fields applied to A and X respectively.

A theoretical account of relaxation in the rotating frame has been provided by Jones [23]. We propose here an approach based on an extension of longitudinal relaxation calculations. The problem is handled by a first transformation, characterized by the operator $\hat{T}$, which removes the time dependence from the hamiltonian. $\hat{T} = \exp[-i\omega t(\hat{I}_z^A + \hat{I}_z^X)]$ in the homonuclear case and $\hat{T} = \exp - i[\omega'_A t\hat{I}_z^A + \omega'_x t\hat{I}_z^X]$ in the heteronuclear case. With $\hat{H}' = \hat{T}\hat{H}\hat{T}^{-1}$, and according to the inequalities given above, this leads to: $\hat{H}' \simeq -\omega_1(\hat{I}_x^A + \hat{I}_x^X)$. This transformation has to be applied as well to the density operator before assessing its evolution under relaxation processes. Likewise, $\hat{H}_0$ has to be replaced by $\hat{H}'$ in the evolution equation. Recall, also, that in the absence of an rf field, the frequency dependence of spectral densities originates from quantities of the type $\exp(i\hat{H}_0 t)\hat{A}_r^m \exp(-i\hat{H}_0 t)$. In the present situation,

Table 2. Frequency dependence of spectral densities which have to be substituted to $\mathcal{J}(0)$ in the presence of a spin-lock field

$\hat{A}_r^m$	
$\hat{I}_x^A$ or $\hat{I}_x^X$	$\mathcal{J}(0)$
$\hat{I}_z^A$ or $\hat{I}_y^A$	$\mathcal{J}(\omega_1)$
$\hat{I}_z^X$ or $\hat{I}_y^X$	$\mathcal{J}(\omega_1)$
$\hat{I}_x^A\hat{I}_x^X$	$\mathcal{J}(0)$
$\hat{I}_z^A\hat{I}_z^X$ or $\hat{I}_y^A\hat{I}_y^X$	$\mathcal{J}(2\omega_1)$

we shall instead be dealing with: $\exp(i\hat{H}'t)\hat{T}^{-1}\hat{A}_r^m\hat{T}\exp(-iH't)$. In the exponential operators, the terms involving $\hat{I}_z$ operators dominate in both cases (without or with an applied rf field), because $\omega_1 \ll (\omega_A, \omega_X$ or $\omega)$ as long as they do not lead to a frequency close to zero. Therefore, $\mathscr{J}(\omega_A)$, $\mathscr{J}(\omega_A + \omega_X)$, in all cases and $\mathscr{J}(\omega_A - \omega_X)$, in the heteronuclear case, are unchanged, but $\mathscr{J}(0)$ must be reconsidered. Depending on the $\hat{A}_r^m$ leading to a non-zero result in Eq. (4), $\mathscr{J}(0)$ has to be changed according to Table 2, which has been set up by noticing that the x axis of the rotating frame plays the same role as the z axis, when dealing with relaxation in the laboratory frame; likewise ω_1 replaces the resonance frequency.

We are then in position to evaluate the evolution of the quantities of interest, namely $\langle \hat{I}_x^A \rangle$ and $\langle \hat{I}_x^X \rangle$, by means of Eq. (5). These two quantities are found to be coupled in a manner analogous to longitudinal magnetizations (Solomon equations):

$$\begin{aligned} d\langle \hat{I}_x^A \rangle/dt &= -R_{1\rho}^A \langle I_x^A \rangle - \sigma_\rho^{A,X} \langle I_x^X \rangle \\ d\langle I_x^X \rangle/dt &= -R_{1\rho}^X \langle I_x^X \rangle - \sigma_\rho^{A,X} \langle I_x^A \rangle \end{aligned} \tag{50}$$

For the homonuclear case, we shall note as usual $\omega_0 = \omega_A = \omega_X$. $R_{1\rho}$ can be decomposed into three contributions: random field, chemical shift anisotropy and dipolar. We obtain:

$$(R_{1\rho}^A)^{rf} = \mathscr{J}^{rf(A)}(\omega_1) + \mathscr{J}^{rf(A)}(\omega_A) \tag{51}$$

$$(R_{1\rho}^A)^{CSA} = \mathscr{J}^{CSA(A)}(\omega_1) + (3/4)\mathscr{J}^{CSA(A)}(\omega_A) \tag{52}$$

which are valid for both homonuclear or heteronuclear cases.

Concerning dipolar contributions, the following expressions hold:

$$\begin{aligned} (R_{1\rho}^A)^d = {} & (1/20)(\mu_0/4\pi)^2(\gamma_A\gamma_X\hbar/r_{AX}^3)^2[2\tilde{J}(2\omega_1) + \tilde{J}(\omega_A - \omega_X)/2 + 3\tilde{J}(\omega_A)/2 \\ & + 3\tilde{J}(\omega_X) + 3\tilde{J}(\omega_A + \omega_X)] \end{aligned} \tag{53}$$

valid for the heteronuclear case and identical to R_2^A when ω_1 becomes vanishingly small (see Eq. (38)).

In the homonuclear case, this becomes:

$$(R_{1\rho}^A)^d = (R_{1\rho}^X)^d = (1/20)(\mu_0/4\pi)^2(\gamma^2\hbar/r_{AX}^3)^2[5\tilde{J}(2\omega_1)/2 + 9\tilde{J}(\omega_0)/2 + 3\tilde{J}(2\omega_0)] \tag{54}$$

Regarding the cross-relaxation terms, one has for a heteronuclear system:

$$\begin{aligned} \sigma_\rho^{XA} = {} & (1/20)(\mu_0/4\pi)^2(\gamma_A\gamma_X\hbar/r_{AX}^3)^2[\tilde{J}(2\omega_1) + \tilde{J}(\omega_A - \omega_X) \\ & + 3\tilde{J}(\omega_A)/2 + 3\tilde{J}(\omega_X)/2] \end{aligned} \tag{55}$$

In the homonuclear case, the tranverse-cross relaxation term is of great use in the CAMELSPIN or ROESY two-dimensional experiments [24, 25]:

$$\mathcal{J}^{Q}(\omega) = (3\pi^2/20)[1/I^2(2I-1)]\chi^2(1+\eta_Q^2/3)\tilde{J}(\omega) \tag{60}$$

Proceeding further in this analysis, one would find coupling between the two quantities $2\langle \hat{I}_z^A \hat{I}_x^X \rangle$ and $2\langle \hat{I}_x^A \hat{I}_z^X \rangle$ which represent A and X doublets, antiphase along the x axis of the rotating frame. This coupling originates again from the CSA–dipolar interference terms (as in the case of longitudinal relaxation). This leads to four simultaneous differential equations whose consideration is however beyond the scope of this chapter.

2.3.6 Quadrupolar Relaxation

Application of the general evolution equation (5) to $\langle \hat{I}_z \rangle$ for a nucleus of spin $I > 1/2$, subjected essentially to quadrupolar interaction, does not lead, in the general case and with the exception of $I = 1$, to a monoexponential recovery [6], because $\hat{I}_z$ is not systematically retrieved after the calculation of commutators involved in (5). In fact, a $(2I-1)$-fold multiexponentiality prevails. A similar situation holds for $\langle \hat{I}_x \rangle$ and $\langle \hat{I}_y \rangle$. A first exception to this statement concerns the extreme narrowing situation, for which a true monoexponential recovery can be achieved for both longitudinal and transverse magnetizations with relaxation rates given by:

$$R_1^Q = R_2^Q = \{(3\pi^2/10)(2I+3)/[I^2(2I-1)]\}(e^2qQ/h)^2(1+\eta_Q^2/3)\tau_c \tag{57}$$

e^2qQ/h is the quadrupole coupling constant expressed in Hz (Q: nuclear quadrupole moment, q: the largest electric field gradient value, η_Q: the asymmetry parameter as defined in Sect. 2.2). This situation is anyway irrelevant in the present context since no variation with B_0 can be expected.

Outside the extreme narrowing regime, and when an apparent monoexponential recovery is observed, it is possible to rely on approximate relaxation rates whose expressions have been derived by Halle and Wennerström [26] by means of perturbation calculations:

$$(R_1^Q)^{app} = [(2I+3)/5][2\mathcal{J}^Q(\omega_0) + 8\mathcal{J}^Q(\omega_0)] \tag{58}$$

$$(R_2^Q)^{app} = [(2I+3)/5][3\mathcal{J}^Q(0) + 5\mathcal{J}^Q(\omega_0) + 2\mathcal{J}^Q(2\omega_0)] \tag{59}$$

where $\mathcal{J}^Q$ proceeds from the general definition of spectral densities (Eq. and arises solely from rotational motion. In the case of an isotropic motion, $\mathcal{J}^Q(\omega)$ can be expressed as a function of the usual reduced spectral density:

$$\mathcal{J}^{Q}(\omega) = (3\pi^2/20)[1/I^2(2I-1)]\chi^2(1+\eta_Q^2/3)\tilde{J}(\omega) \tag{60}$$

with

$$\chi = eQV_{zz}/h \tag{61}$$

V_{zz} is the largest element of the field gradient tensor, expressed in its principal axis system. Equation (60) also applies to an axially symmetric tensor with, of course, $\eta_Q = 0$, regardless of the nature of the reorientational motion. It can be noticed that Eqs. (58) to (61) rigorously apply to nuclei of spin $I = 1$, for which monoexponentiality always holds [6]. The special case of nuclei of spin $I = 3/2$ is worth mentioning since analytical solutions can be derived [27]. They have been obtained here with the help of Redfield formulas. The biexponentiality of longitudinal recovery is defined by two rates R'_1 and R''_1 of respective weights 4/5 and 1/5:

$$R'_1 = 12\mathscr{J}^Q(2\omega_0) \qquad (62)$$

$$R''_1 = 12\mathscr{J}^Q(\omega_0)$$

Concerning transverse relaxation, two rates again occur, R'_2 and R''_2 with respective weights 3/5 and 2/5:

$$R'_2 = 6[\mathscr{J}^Q(0) + \mathscr{J}^Q(\omega_0)] \qquad (63)$$

$$R''_2 = 6[\mathscr{J}^Q(\omega_0) + \mathscr{J}^Q(2\omega_0)]$$

These relaxation rates, in appropriate conditions of isotropic motion or of axially symmetric tensor, can again be expressed in terms of reduced spectral densities (see Eq. (60)):

$$\begin{aligned} R'_1 &= (\pi^2/5)\chi^2(\eta^2 + 3)\tilde{J}(2\omega_0) \\ R''_1 &= (\pi^2/5)\chi^2(\eta^2 + 3)\tilde{J}(\omega_0) \end{aligned} \qquad (62\text{ bis})$$

$$\begin{aligned} R'_2 &= (\pi^2/10)\chi^2(\eta^2 + 3)[\tilde{J}(0) + \tilde{J}(\omega_0)] \\ R''_2 &= (\pi^2/10)\chi^2(\eta^2 + 3)[\tilde{J}(\omega_0) + \tilde{J}(2\omega_0)] \end{aligned} \qquad (63\text{ bis})$$

It can be interestingly noticed that $(4/5)R'_1 + (1/5)R''_1$ and $(3/5)R'_2 + (2/5)R''_2$ which would correspond to initial slopes of recovery curves, are identical to $(R_1^Q)^{app}$ and $(R_2^Q)^{app}$ (Eqs. (58) and (59)). Finally, the lineshape will reflect the existence of two transverse relaxation rates and will appear as the superposition of two Lorentzian lines of respective weights 3/5 and 2/5, and of width of half-height R'_2/π and R''_2/π respectively [28]. These two lines correspond to the $-1/2 \rightarrow +1/2$ transition (of weight 3/5), and to the $-3/2 \rightarrow -1/2$ and $1/2 \rightarrow 3/2$ transitions (of weight 2/5), respectively.

Inspection of the above formulas reveals for longitudinal relaxation a B_0 dependence similar to the one found for dipolar relaxation. Conversely, as far as transverse relaxation is concerned, not only linewidths but also lineshapes are expected to exhibit strong variations with B_0. This is examplified by the simulations in Fig. 3 which show the dramatic modification of the lineshape with B_0 for a 3/2 spin nucleus. The simulations in Fig. 3 have been performed with realistic parameters of ^{23}Na [28] with a reduced spectral density of the form $2\tau_c/(1 + \omega^2\tau_c^2)$.

Another phenomenon may alter lineshapes: the so-called *dynamic frequency shift*. Although this is a very general feature, originating from the imaginary

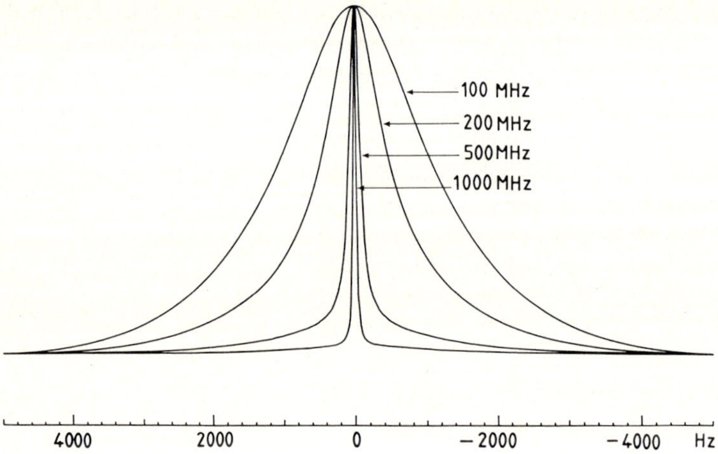

Fig. 3. Evolution with B_0 of ^{23}Na ($I = 3/2$) lineshape. Quadrupolar coupling: 0.5 MHz, correlation time: 5×10^{-9} s. The corresponding ^1H resonance frequencies are given on the figure

part of the spectral densities, noticeable manifestations are expected only in the case of quadrupolar transverse relaxation [29], [30]. In order to delineate the essential properties of dynamic frequency shifts, let us consider the simple case of a reduced spectral density with a monoexponential correlation function:

$$\tilde{J}(\omega) = \frac{1}{4\pi}\int_{-\infty}^{+\infty} \overline{Y_2^0(t)Y_2^0(0)}\exp(-i\omega t)\,dt,$$

which, with $\overline{Y_2^0(t)Y_2^0(0)} = |Y_2^0|^2 \exp - |t|/\tau_c$, leads to

$$\tilde{J}(\omega) = \int_{-\infty}^{+\infty} \exp(-|t|/\tau_c)\exp(-i\omega t)\,dt$$

and upon complete integration:

$$\tilde{J}(\omega) = 2\tau_c/(1 + \omega^2\tau_c^2) - i(2\omega\tau_c^2)/(1 + \omega^2\tau_c^2) \tag{64}$$

The first term in the right hand-side of (64) is of course the damping term which has been discussed throughout. The imaginary part is almost always neglected because of the usual weakness of $\omega\tau_c^2$. Since it is imaginary, it is *not a damping term but a frequency term* which should be included in the precession part of the general evolution equations (5) and (9). However, this contribution is non-negligible only in the limit of an appreciable value of $\omega\tau_c^2$. In such a case, $\omega^2\tau_c^2$ will be much larger than 1 and the dynamic frequency shift $\Delta\nu_d$ can be approximated by:

$$\Delta\nu_d \simeq 2/\omega \tag{65}$$

An immediate consequence of this result is *the decrease of the dynamic frequency shift when* B_0 *increases.* Of course, a spectral density $\tilde{J}(0)$ or $\mathscr{J}(0)$ does not yield any frequency shift. These properties can be illustrated by the transverse relaxation of a nucleus with spin $I = 3/2$ (vide supra). The two lorentzian lines of respective weights 3/5 and 2/5 do not entail the same dynamic shift, with, of course, the prerequisite of a $\omega\tau_c^2$ value large enough. Shifts proportional to $2/\omega_0$ and $3/\omega_0$ arise from R'_2 and R''_2, respectively and the positions of the maxima of the two associated lines differ by a quantity of the order of χ^2/ω_0 (see (63a)). Consequently, an appreciable effect cannot be expected unless the square of the quadrupole coupling constant is of the same order as the measurement frequency. In that case, and especially at high fields, the importance of linewidths would preclude its detection.

2.3.7 Scalar Relaxation of the Second Kind

This mechanism operates on a slowly relaxing nucleus A, itself J-coupled to a fast relaxing nucleus X (generally a quadrupolar nucleus) [Ref. 6 p. 309]. Viewed from A, the fast relaxation of X plays a role similar to that of molecular motion through a time modulation of the indirect coupling interaction. The contributions R_1^{SC} and R_2^{SC} to longitudinal and transverse relaxation rates may vary with the applied field via the frequency dependence of T_1^X and T_2^X. It can be noticed from Table 1 that the spin operators $\hat{A}_r^m$ are identical to those pertaining to random fields. Therefore, from Sect. 2.3.1:

$$R_1^{SC} = 2\mathscr{J}_1^{SC}(\omega_A) \tag{66}$$

$$R_2^{SC} = \mathscr{J}_0^{SC}(0) + \mathscr{J}_1^{SC}(\omega_A) \tag{67}$$

The general properties derived by Hubbard concerning the independence of spectral densities with respect to the projection m, and which arise from the isotropy of the medium, no longer apply here because of the special form of the "space functions" F_r^m (see Table 1). The projection must therefore be specified for each spectral density and one obtains (J_{AX} being the coupling constant between A and X):

$$\begin{aligned} \mathscr{J}_0^{SC}(0) &= J_{AX}^2 \int_0^\infty \overline{\hat{I}_z^X(t)\hat{I}_z^X(0)}\,dt \\ \mathscr{J}_1^{SC}(\omega_A) &= J_{AX}^2 \int_0^\infty \overline{\hat{I}_+^X(t)\hat{I}_-^X(0)}\exp(-i\omega_A t)\,dt \end{aligned} \tag{68}$$

If it is assumed that the longitudinal and transverse X magnetizations evolve exponentially according to their specific relaxation times T_1^X and T_2^X, the following relationships hold:

$$\overline{\hat{I}_z^X(t)\hat{I}_z^X(0)} = |\hat{I}_z^X(0)|^2 \exp(-t/T_1^X) = [I_X(I_X+1)/3]\exp(-t/T_1^X)$$

$$\begin{aligned} \overline{\hat{I}_+^X(t)\hat{I}_-^X(0)} &= \overline{\hat{I}_+^X(0)\hat{I}_-^X(0)}\exp(i\omega_X t)\exp(-t/T_2^X) \\ &= [2I_X(I_X+1)/3]\exp(i\omega_X t)\exp(-t/T_2^X) \end{aligned}$$

The latter equation arises from the identity; $\hat{I}_+\hat{I}_- = \hat{I}_x^2 + \hat{I}_y^2 - i\hat{I}_z$ with $\overline{\hat{I}_z} = 0$. Inserting these results in the expression of spectral densities given above yields:

$$R_1^{SC} = [2J_{AX}^2 I_X(I_X+1)/3]T_2^X/[1+(\omega_A-\omega_X)^2(T_2^X)^2] \quad (66\text{ bis})$$

$$R_2^{SC} = [J_{AX}^2 I_X(I_X+1)/3]\{T_1^X + T_2^X/[1+(\omega_A-\omega_X)^2(T_2^X)^2]\} \quad (67\text{ bis})$$

It can be recalled that in the above formulas J_{AX} *is expressed in rad. s*$^{-1}$. Also, as is well known, scalar relaxation of the second kind affects essentially R_2 unless frequencies of A and X are sufficiently close to each other. In that latter case, both R_1 and R_2 are subjected to this mechanism.

3 A Survey of Molecular Motion Models in Liquids

To establish the mathematical expressions of the spectral density, of the relaxation times T_1, T_2, $T_{1\rho}$ and of the NOE, one needs a model which describes the molecular motions. The problem of the rotational and translational molecular motions has been discussed in several review articles [31], [32]. We shall give here a brief survey of the models which have been applied in NMR studies and we shall limit ourselves to molecules of low molecular weight, with M equal to a few hundreds Dalton. Two kinds of motion must be considered, the translational and rotational ones respectively.

3.1 Isotropic Translational and Rotational Diffusion

3.1.1 Translational Diffusion

In a very first approach, a molecule may be considered as a rigid sphere of radius a, which moves isotropically in a medium of viscosity η, the molecular translational diffusion may be described by a random walk model. That is, starting from a given origin the molecule is displaced by a vector $\vec{l}_1$ in time t_1, at this point its previous history is forgotten and the molecule is displaced by a vector $\vec{l}_2$ again in a time t_2 and so on indefinitely. Denoting by $\vec{r}$ the vector which characterizes the molecular position at time t, the probability density function $\psi(\vec{r},t)$ of finding the molecule at position $(\vec{r},t)$ if it was initially at position $\vec{r}=0$, satisfies a diffusion equation [33]:

$$\frac{\partial\psi(\vec{r},t)}{\partial t} = D\Delta\psi(\vec{r},t) \quad (69)$$

D the diffusion constant is equal to $\overline{l_1^2}/6t_1$ with:

$$\overline{l_1^2} = \int l_1^2 \Psi(l_1,t_1)\,d^3l_1$$

and may be shown to be independent of t_1. The diffusion constant D is related to the viscosity by $D = kT/6\pi\eta a$. The solution of equation (69) is given by:

$$\psi(\vec{r}, t) = \frac{1}{(4\pi Dt)^{3/2}} \exp - \left(\frac{r^2}{4Dt} \right) \tag{70}$$

If one considers two identical spherical molecules, which are defined by a relative distance r_0 at time $t = 0$, the probability density function to have these molecules a distance r apart at time t, is simply obtained from Eq. (70) by replacing D by 2D

$$\psi(\vec{r}, t) = \frac{1}{(8\pi Dt)^{3/2}} \exp - \left[\frac{(r - r_0)^2}{8Dt} \right] \tag{71}$$

From this equation the spectral density function $\mathscr{J}(\omega)$ corresponding to the intermolecular dipole interactions may be obtained [6]. This involves the calculation of the integrals

$$\alpha^{(m)}(8\pi Dt)^{-3/2} \int \frac{Y_1^{m*}(\Omega_0) Y_1^m(\Omega)}{r_0^3} \exp - \frac{(r - r_0)^2}{8Dt} dr_0^3 \, dr^3$$

Where $\alpha^{(m)}$ is a numerical factor which depends on the considered spherical harmonics Y_1^m. Denoting by d the closest distance of the interacting molecules, and by N the number of interacting spins by volume unit it comes:

$$\mathscr{J}(\omega) = \frac{N}{dD} \int_0^\infty (J_{3/2}(u))^2 \frac{u \, du}{u^4 + \omega^2 \tau^2}, \quad \tau = \frac{d^2}{2D}$$

and $J_{3/2}(u)$ is a Bessel function. Using the expression of $J_{3/2}^2(u)$, $\mathscr{J}(\omega)$ may be explicitly calculated. After integration it comes:

$$\mathscr{J}(\omega) = \frac{2}{\pi} \frac{N}{dD} X^{-5/2} [(X - 2) + (X - 2) \sin \sqrt{X} + (X + 4\sqrt{X} + 2) \cos \sqrt{X}] \exp(-\sqrt{X}) \tag{72}$$

where $X = \omega\tau_c$. The behaviour of $\mathscr{J}(X)$ is shown in Fig. 4. For $\omega\tau_c > 4 \times 10^{-3}$ it is a decreasing function, nothing special is expected at high-field.

3.1.2 Rotational Brownian Motion

The random walk model may be applied to the description of the rotational motion of a spherical molecule of radius a. The molecule starting from a given orientation at a given time rotates by an angle θ_1 within a time t_1. At this time its previous history is forgotten and it starts to rotate by an angle θ_2 in a time t_2 and so on indefinitely. The small angular displacement hypothesis also implies that the angular momentum correlation time is short compared to the reorientational correlation time. Such a model is particularly well adapted for atomic

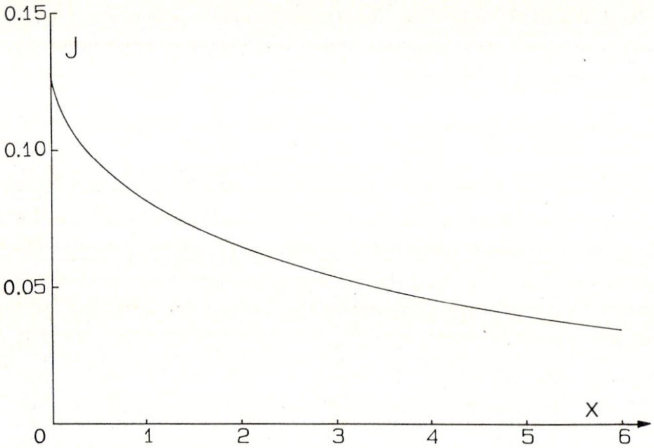

Fig. 4. Plot of the translational density function as a function of $X = 2\pi\nu\tau_c$, ν observation frequency, τ_c correlation time (see Eq. 72)

liquids or macromolecules, but it fails when the molecules are not spherical in shape. The probability density function $\psi(\Omega, t)$ that a given molecular axis, defined by a direction Ω_0 at time $t = 0$, and by Ω at time t, follows a diffusion equation, which may be established as for the random walk translational motion

$$\frac{\partial\psi(\Omega,t)}{\partial t} = \frac{D}{a^2}\Delta_\Omega\psi \tag{73}$$

Δ_Ω is the angular part of the spherical Laplacian operator, a is the molecular sphere radius, D is the rotation diffusion constant $D = kT/8\pi a\eta$ where η is the

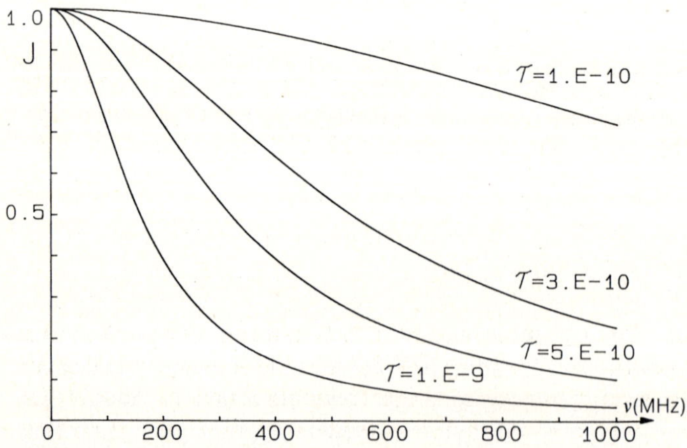

Fig. 5. Variation of the simplest reduced spectral density $\tilde{J}(\omega) = (2\tau_c/(1 + 4\pi^2\nu^2\tau_c^2))$ as a function of the frequency for various correlation times τ_c(s). All the spectral densities have been normalized to 1 for $\nu = 0$

kinematic viscosity coefficient. With the initial condition $\psi(\Omega, 0) = \delta(\Omega - \Omega_0)$, and expanding the Dirac delta function as a sum of spherical harmonics $\delta(\Omega - \Omega_0) - \Sigma_{l,m} Y_l^{m*}(\Omega)$, the solution of Eq. (73) may be obtained:

$$\psi(\Omega, t) = \sum_{l,m} Y_l^{m*}(\Omega_0) Y_l^m(\Omega) \exp - \left(\frac{t}{\tau_1}\right) \tag{74}$$

$$\tau_1 = \frac{a}{l(l+1)D}$$

Since we are essentially dealing with spherical harmonics for rank two, the reduced spectral density $\tilde{J}(\omega)$ is given by $\tilde{J}(\omega) = 2\tau_c/(1 + \omega^2\tau_c^2)$, where τ_c stands for τ_2. The behaviour of the spectral density with the frequency, for different values of the correlation time τ_c, and in the NMR frequency range 100 MHz up to 1 GHz are shown in Fig. 5.

3.1.3 *Extended Diffusion Model*

In the extended diffusion model, the restriction of small angle diffusion steps is removed [34]. In the time between collisions the molecule is assumed to rotate freely and the time over which the torque acts on the molecule is short compared to the time required for a complete rotation of the molecule. The time between collisions is a random variable characterized by its mean τ_J, the correlation time for the rotational angular momentum. The diffusion is characterized by a stochastic process, in which the collision probability follows a Poisson distribution. The probability for a molecule to be in its nth difusive step at time t is given by

$$\Psi(n, t) = \left(\frac{t}{\tau_J}\right)^{n-1} \exp\left(-\frac{t}{\tau_J}\right) \Big/ n - 1 \tag{75}$$

Two extended rotational diffusion processes which differ in the nature of the collisions are considered. In the M diffusion model, EDM, only the orientation of the angular momentum is randomized at the end of each collision. In the J diffusion model, EDJ, both the orientation and the magnitude of the angular momentum are randomized. In a slight generalization of the extended diffusion model, the rate τ_J^{-1} which appears in the Poisson distribution is a function of the molecular rotation frequency.

The extended diffusion model first applied to calculate the reorientational correlation functions of linear molecules was then applied to spherical top molecules [35], to symmetrical top molecules [36] and to asymmetric molecules [37]. Several numerical simulations of the correlation functions and of the spectral densities as a function of the collision frequency τ, have been presented [35]. For a linear molecule, if the correlation time τ is expressed in reduced units $(I/KT)^{1/2}$ (I momentum of inertia), it turns out that, in the EDJ diffusion model, the spectral density function can be approximated by the Lorentzian curve predicted by the Debye model for small values of $\tau(\tau \leqslant 0.1)$. Thus in most cases a discrimination between Debye's model and extended diffusion appears to be difficult by means of NMR relaxation.

3.1.4 *Fokker–Planck–Langevin Model*

The Fokker–Planck–Langevin model (FPL) for the description of the molecular rotation was first introduced by Fixman and Rider [38]. It is an extension of the Debye rotational model, where the molecular rotation is modulated by a retarding torque proportional to its angular velocity and by a rapidly fluctuating Brownian torque, which reflects the molecular nature of the liquid. The physical picture of the EDJ or EDM and FPL models is different. In the EDJ model, the molecules experience intermittent but very strong intermolecular torques. Conversely, in the FPL model the molecules experience intermolecular torques at all times. The conditional probability density $\Psi(\Omega, \omega, t;\ \Omega_0, \omega_0)$, that a molecule has angular velocity $\vec{\omega}$ in $d^3\omega$ and orientation Ω in $d\Omega$ at time t, satisfies a differential equation $\partial\psi/\partial t = Z\psi$ which has been derived in two basic papers [39], [40], where Z is an operator consisting of two parts, Z_s which describes the motion of a freely rotating molecule, and Z_c which describes the effect of molecular collisions. General expressions for the reorientational correlation functions, correlation times, spectral densities, angular correlation function densities have been derived for symmetric top molecules [41] and for asymmetric top molecules [42]. In the last paper, a set of plots of the aforementioned quantities are presented along with a comparison of the values obtained using the EDJ model. For isotropic friction, the two models give almost indistinguishable descriptions of rotational motions in the rotational diffusion limit. The FPL model has been extended to molecules which have an internal rotational degree of freedom [43].

3.2 *Anisotropic Rotational Diffusion*

In two extremely well documented papers [44], [45], F. Perrin extended Debye's work and established the rotational and translation diffusion equations for ellipsoids. Considered as a continous medium, the liquid in which the molecule is immersed exerts on the molecule a retarding force which is proportional to its linear velocity and a retarding torque proportional to its angular velocity. The medium molecular structure is accounted for by linear retarding forces and torques which act in a statistically independent manner during short times Δt. The ellipsoid motion is described by the motion of a point on a hypersphere $\sum$ of dimension four, and the probability density function U having a given point on an elementary surface of $\sum$ is derived. This equation depends on generalized diffusion coefficients for reorientation around three perpendicular axes of the particle. Perrin's analysis of an ellipsoid motion was devoted to the study of physical phenomena, including the dielectric dispersion, depolarisation in fluorescence emission, and then applied to NMR by several authors [46–48].

The equations derived by Perrin have been rewritten in reference [48] when considering the motion of an ellipsoid. Let S' denote a rectangular coordinate system fixed to the ellipsoid and S_0 a fixed system of reference. The probability density P of finding S' at a particular orientation Ω at time t obeys a diffusion law $(\partial P(\Omega, t)/\partial t) = \wedge P(\Omega, t)$. $\wedge$ is a complex second order operator given in Ref. [48]. $\wedge$ depends upon three rotational diffusion coefficients. The average value at time t of any function f which depends on the orientation of the ellipsoid satisfies the equation $(d\langle f\rangle/dt) = \langle \wedge f\rangle$. In this frame a correlation function $\langle f(t)f^*(t+\tau)\rangle$ may be written as $\sum_i C_i . \exp(|\tau|/\tau_i)$ which depends on five correlation times, which could be obtained from the analysis of the molecular motion. As an illustration of the evaluation of a spectral density, let us consider the case of an anisotropic rotational diffusion process involving, for simplicity, only two different diffusion coefficients, namely D_z and $D_x = D_y$ (a model example is a symmetric top molecule). Let us denote by θ_r and φ_r the polar angles of the relaxation vector pertaining to mechanism r in the molecular frame (x, y, z). Calculations are easily carried out according to a procedure devised by Hubbard [13], or alternatively by following Huntress [31]. Both approaches yield the same result. With $1/\tau_1 = 2D_x + 4D_z$; $1/\tau_2 = 5D_x + D_z$ and $1/\tau_3 = 6D_x$, the following result is obtained for a cross-correlation reduced spectral density:

$$\begin{aligned}\tilde{J}^{rr'}(\omega) = {} & (3/4)\sin^2\theta_r \sin^2\theta_{r'} \cos 2(\varphi_r - \varphi_{r'})[2\tau_1/(1+\omega^2\tau_1^2)] \\ & + (3/4)\sin(2\theta_r)\sin(2\theta_{r'})\cos(\varphi_r - \varphi_{r'})[2\tau_2/(1+\omega^2\tau_2^2)] \\ & + (1/4)(3\cos^2\theta_r - 1)(3\cos^2\theta_{r'} - 1)[2\tau_3/(1+\omega^2\tau_3^2)]\end{aligned} \tag{76}$$

Of course, the autocorrelation spectral density $\tilde{J}(\omega)$ is deduced from the above expression by equating θ_r and $\theta_{r'}$, on the one hand, and φ_r and $\varphi_{r'}$ on the other hand:

$$\begin{aligned}\tilde{J}(\omega) = {} & (3/4)\sin^4\theta_r[2\tau_1/(1+\omega^2\tau_1^2)] + (3/4)\sin^2(2\theta_r)[2\tau_2/(1+\omega^2\tau_2^2)] \\ & + (1/4)(3\cos^2\theta_r - 1)^2[2\tau_3/(1+\omega^2\tau_3^2)]\end{aligned} \tag{77}$$

$\tilde{J}(\omega)$ exhibits a complicated behaviour with respect to the frequency. For the peculiar case where $D_x \ll D_z$, only τ_3 depends of the slow motion, and $\tilde{J}(\omega)$ reduces to the first two contributions. The slow motion influence may also disappears for $(3\cos^2\theta_r - 1) = 0$.

In macromolecules or in aggregates such as micelles the description of the motion becomes more complicated as its exists an overall motion of the molecule and internal motions. One may for example consider the reorientation of an inter-nuclear vector AB. Owing to the intrachain steric interactions the AB vector does not assume all possible orientations. For the special case where the overall motion can be described by a single correlation time τ_m, the orientation of the AB vector involves three parameters [49], namely a generalized order parameter S which is the measure of the degree of the spatial restriction of the local motion, an effective correlation time τ_e for the internal motion and τ_m. In such a case the reduced spectral density $\tilde{J}(\omega)$ may be written

$$\tilde{J}(\omega) = \left[\frac{2S^2\tau_m}{1+\omega^2\tau_m^2} + \frac{2(1-S^2)\tau_2}{1+\omega^2\tau_2^2}\right],$$

with $\tau_2^{-1} = \tau_m^{-1} + \tau_e^{-1}$. If the overall motion is considerably smaller than the internal motion ($\tau_m \gg \tau_e$), one may reach a situation where the internal motion is in the extreme narrowing limit, but not the overall motion. Thus it comes:

$$\tilde{J}(\omega) = \left(\frac{2S^2\tau_m}{1+\omega^2\tau_m^2} + 2(1-S^2)\tau_e\right) \tag{78}$$

The spin relaxation time T_2 will be dependent of both τ_m and τ_e via $\tilde{J}(0)$. However at high frequency the influence of τ_m on T_1 may be cancelled, due to the large value of $\omega^2\tau_m^2$.

4 Sensitivity and Relaxation Rate

We shall consider the sensitivity of an NMR Fourier Transform experiment, taking mainly into account the effects of the relaxation parameters on the signal-to-noise ratio (S/N), which is often a quantity of major interest to the spectroscopist. In the general discussion, we leave apart the instrumental features, which are supposed to remain identical with frequency. In the frequency domain of interest, the temperature dependence through the Boltzmann distribution has no effect. We shall further assume that a large number of scans is performed, so that the flip angle α of each scan can be optimized according to the Ernst value [50]:

$$\cos\alpha = \exp(-T/T_1) = E_1 \tag{79}$$

Where T is the recycle time, i.e. the time elapsed between two consecutive pulses. The magnetization loss, due to non complete return to thermal equilibrium, prior to each new pulse depends on E_1 and the collected signal is proportional to:

$$[(1-E_1)/(1+E_1)]^{1/2} \tag{80}$$

The S/N enhancement is proportional to the square root of the number of scans and therefore to $(T)^{-1/2}$. The other factors which govern the sensitivity and which do not depend of the recycle time T are: i) $B_0^{3/2}$ which is assumed to represent the increase of the signal intensity, arising from nuclear magnetization and from the detection scheme. The exponent of 3/2, whose validity has been questioned [51], has not to be considered too strictly, it merely represents a basis for the present discussion, which is essentially devoted to the effect of relaxation; ii) T_2, which denotes the effective transverse relaxation time, since the absorption signal amplitude is proportional to T_2 according to the

Bloch equations; iii) $(1+\eta)$ which accounts for the Nuclear Overhauser Effect, in cases where the nucleus under study is observed with irradiation of the other spins to which it is coupled through dipolar interaction. Altogether we arrive at:

$$S/N \propto B_0^{3/2}(1+\eta)T_2[(1-E_1)/(1+E_1)]^{1/2}(T)^{-1/2} \tag{81}$$

We have still to choose an appropriate value for T. It turns out that (81) is maximum for $T=0$, which is of course unrealistic since we must avoid distortions due to free induction decay truncation. $T=3T_2$ is recognized to be a good compromise. This yields

$$S/N \propto B_0^{3/2}(1+\eta)(T_2/3)^{1/2}[(1-E)/(1+E)]^{1/2}$$
$$\text{with } E=\exp(-3T_2/T_1) \tag{82}$$

It turns out that S/N is generally a complicated function of B_0 via the frequency dependence of the relaxation parameters. However, in the extreme narrowing limit, a $B_0^{3/2}$ dependence should be observed. Another limiting case concerns the situation where $T_2 \ll T_1$. In that case (82) reduces to

$$S/N \propto B_0^{3/2}(1+\eta)(T_2/2T_1)^{1/2} \tag{83}$$

In general (83) also shows a complex dependence in B_0. However, it must be pointed out that if T_2 has reached its asymptotic value, although T_1 still increases with frequency, the S/N gain with B_0 will be less than the usual $B_0^{3/2}$ dependence. Further discussion of the S/N with the magnetic field be given in the next section.

5 Behaviour at High Field of Relaxation Parameters and Sensitivity of Most Common Nuclei

This section takes representative nuclei one by one (^1H, ^{13}C, ^{15}N, ^{31}P) and shows the variation of dynamical parameters and sensitivity with the applied field B_0. The relaxation parameters are the longitudinal relaxation time T_1, the transverse relaxation time T_2, the cross-relaxation terms σ and σ_ρ and the related NOE enhancement factor η. The field varies from 4.7 T up to 23.5 T (200 MHz $\leqslant \nu(^1H) \leqslant$ 1000 MHz). We label the fields by their equivalent hydrogen resonance frequency, as is customary for the NMR spectroscopist.

We shall restrict ourselves to intramolecular dipolar and CSA mechanisms and to the simplest available reduced spectral density (see Sect. 3):

$$\tilde{J}(\omega)=\frac{2\tau_c}{1+\omega^2\tau_c^2}$$

where τ_c is an effective correlation time. The variation of $\tilde{J}(\omega)$ in the frequency range of interest and for different values of τ_c is shown in Fig. 5. Referring to

Table 3. Correlation time τ_c (ns) for which $\omega\tau_c = 1$ for different nuclei at a field of 11.4 T ($\nu(^1H) = 500$ MHz)

Nucleus	^1H	^{31}P	^{13}C	Pt	^2H	^{15}N
τ_c(ns)	0.32	0.79	1.27	1.48	2.07	3.14
ν(MHz)	500	202.4	125.1	107.5	76.5	50.7

Sect. 2 and considering relaxation of spin A, we notice that i) homonuclear dipolar interaction, as well as quadrupolar relaxation, involves $\tilde{J}(\omega_A)$ and $\tilde{J}(2\omega_A)$; ii) heteronuclear dipolar interaction between A and X involves $\tilde{J}(\omega_A)$, $\tilde{J}(\omega_A + \omega_X)$ and $\tilde{J}(\omega_A - \omega_X)$; iii) CSA mechanism involves $\tilde{J}(\omega_A)$. Relaxation parameters become really frequency dependent and therefore sensitive to a given molecular motion, whenever the condition $\omega\tau_c = 1$ is satisfied around the considered frequency. Relevant values of τ_c, regarding the spectral densities given above, are reported in Table 3. Also it appears that the smallest correlation time producing a detectable variation of spectral densities is $8\ 10^{-11}$ s (^1H observed up to 1 GHz).

5.1 *Hydrogen (^1H)*

Proton chemical shift is generally small ($\Delta\sigma \leqslant 20$ ppm [52]). It reaches its highest value in the case of hydrogen bonds. The dipolar coupling is strong, because of its high gyromagnetic ratio and the proximity to other nuclei, in particular other protons. Thus homonuclear dipolar contribution to relaxation is predominant and we report in Table 4 the evolution of T_1 (like) and T_2 (like),

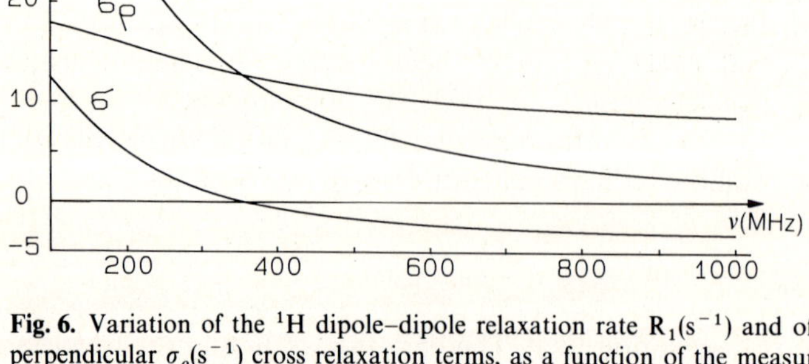

Fig. 6. Variation of the ^1H dipole–dipole relaxation rate $R_1(s^{-1})$ and of the parallel $\sigma(s^{-1})$ and perpendicular $\sigma_\rho(s^{-1})$ cross relaxation terms, as a function of the measurement frequency ν for a correlation time of 5×10^{-10} s, $r_{H \cdots H} = 1.4$ Å

Table 4. Evolution of the proton–proton relaxation times T_1 and T_2 is ms as a function of the frequency, for different values of the correlation times τ_c(s); $r_{H \cdots H} = 1.4$ Å. s is a quantity proportional to the signal to noise ratio S/N (Eq. (81) without NOE, $\eta = 0$). s is normalized to 1 at 200 MHz

ν_0(MHz)		200	400	600	800	1000
$\left(\frac{B_0}{B_{200}}\right)^{3/2}$		1	2.83	5.20	8	11.18
$\tau_c = 5 \times 10^{-9}$	T_1	177	699	1568	2786	4357
	T_2	5.6	5.8	5.8	5.9	5.9
	s	1	1.48	1.83	2.12	2.37
$\tau_c = 5 \times 10^{-10}$	T_1	39	94	183	305	462
	T_2	24	34	42	47	50
	s	1	3.07	5.44	7.61	9.46
$\tau_c = 5 \times 10^{-11}$	T_1	179	186	197	213	234
	T_2	177	180	184	190	196
	s	1	2.85	5.31	8.30	11.85

corresponding to the interaction of a pair of protons (distant by (1.4 Å). Also in this table is given a parameter which allows to compare the actual sensitivity increase (Eq. (82)) with $B_0^{3/2}$. The main conclusion is that the $B_0^{3/2}$ factor is only reached when τ_c is sufficiently small.

The homonuclear Overhauser effect and its associated cross-relaxation parameter σ are of special importance, as they play a central role in determining non-bonded geometrical distances in biologically significant molecules or ensembles (two-dimensional NOE spectroscopy—NOESY). Its r^{-6} dependence and the conditions for its measurement are discussed in Chap. 7 of this book and e.g. in Ref. [53], Sect. 9.7. We have thus plotted, in Fig. 6, R_1 (like) and σ vs. ν, the resonance frequency, for a pair of interacting protons distant by 1.4 Å and for a correlation time of 5×10^{-10} s, a value which is realistic for biological molecules of moderate size. As pointed out several times σ becomes small or even zero around the mid-point of the considered frequency range. This leads to make use of σ_ρ, the cross-relaxation term in the rotating frame (sometimes called also σ perpendicular, by difference to the conventional σ, called σ parallel). Measured in the presence of a spin locking radiofrequency field (see Sect. 2.3.5), σ_ρ is the basis of the two-dimensional ROESY experiments [24, 25]. σ_ρ has also been plotted in Fig 6; this plot exemplifies the advantages of ROESY over NOESY in certain frequency domains.

5.2 *Carbon* ^{13}C

The interest of very high fields for ^{13}C NMR spectroscopy lies primarily in the gain in sensitivity provided by the expected $B_0^{3/2}$ enhancing factor. It is thus interesting to examine if this gain in sensitivity is offset by the shortening of

T_1, which lengthens the time between pulses in multiscan experiments. Tables 5a, 5b and 5c show the variation of the relaxation times T_1 and T_2, of the NOE enhancement factor η and of the sensitivity with the frequency for different kinds of ^{13}C: aliphatic $\left(C\text{–}C{<}^{H}_{H}\right)$, olefinic $\left(>C\text{=}C{<}^{H}\right)$ and carboxylic $\left(^{H}>C\text{=}O\right)$. In each case realistic values for the magnetic shielding tensor anisotropy $\Delta\sigma$ [14, 52, 54] and for the C···H distances have been considered. The three correlation time values 5×10^{-11} s, 5×10^{-10} s and 5×10^{-9} s respectively correspond to small organic molecules and to moderately large biochemical molecules. The evolution of the relaxation times with the frequency are shown on Figs. 7 and 8. For the carbonyl group, owing to the large magnetic shielding anisotropy, and due to the absence of NOE enhancement, one observes a decrease in sensitivity with the applied magnetic field. For the aliphatic and olefinic carbons, the situation is much more favorable. In a Log–Log diagram, the sensitivity may be reasonably well approximated by a linear variation, $\text{Log}\, s = n \,\text{Log}\, B_0 + a$. The values of the coefficients n are shown in Table 6, for different correlation times. Generally the sensitivity increase with the field will be less than $B_0^{3/2}$ i.e. $n < 1.5$. However, in some situations there is a marked increase of sensitivity (with respect to $B_0^{3/2}$), due to the evolution of relaxation

Table 5a. Aliphatic $^{13}C\left(C\text{–}C{<}^{H}_{H}\right)$ $r_{C\text{–}H} = 1.09$ Å; $\Delta\sigma = 25$ ppm. T_1 spin–lattice relaxation time; T_2 spin–spin relaxation time; η NOE factor (Eq. 33); s_0 sensitivity without NOE, normalized to 1 at 200 MHz, s sensitivity with NOE, normalized to 1 at 200 MHz; τ_c correlation time (s)

ν_0(MHz)		200	400	600	800	1000
$\left(\frac{B_0}{B_{200}}\right)^{3/2}$		1	2.83	5.20	8	11.18
$\tau_c = 5 \times 10^{-9}$	T_1 (ms)	47	151	323	562	866
	T_2 (ms)	18	21	22	23	23
	η	0.21	0.17	0.16	0.16	0.15
	s_0	1	1,95	2.56	3.02	3.41
	s	1	1.88	2.45	2.89	3.26
$\tau_c = 5 \times 10^{-10}$	T_1 (ms)	62	93	125	159	194
	T_2 (ms)	59	82	103	120	133
	η	1.54	0.95	0.62	0.45	0.35
	s_0	1	3.29	6.67	10.86	15.61
	s	1	2.53	4.26	6.19	8.30
$\tau_c = 5 \times 10^{-11}$	T_1 (ms)	467	473	481	493	508
	T_2 (ms)	467	471	478	487	498
	η	1.98	1.96	1.93	1.89	1.84
	s_0	1	2.84	5.25	8.15	11.51
	s	1	2.82	5.16	7.91	11.97

Table 5b. Olefinic ^{13}C $\left(>C{=}C<^{H}\right)$ $r_{C\text{–}H} = 1.09$ Å; $\Delta\sigma = 200$ ppm. $T_1, T_2, \eta, s_o, s, \tau_c$: see Table 5a

ν_0(MHz)		200	400	600	800	1000
$\left(\frac{B_0}{B_{200}}\right)^{3/2}$		1	2.83	5.20	8	11.18
$\tau_c = 5 \times 10^{-9}$	T_1 (ms)	87	234	390	520	618
	T_2 (ms)	34	32	26	20	15
	η	0.19	0.130	0.10	0.07	0.05
	s_0	1	1.74	1.98	2.02	1.97
	s	1	1.65	1.82	1.81	1.74
$\tau_c = 5 \times 10^{-10}$	T_1 (ms)	121	158	169	162	151
	T_2 (ms)	114	138	135	117	98
	η	1.49	0.81	0.42	0.23	0.14
	s_0	1	3.06	5.47	7.68	9.53
	s	1	2.22	3.11	3.79	4.34
$\tau_c = 5 \times 10^{-11}$	T_1 (ms)	912	860	784	697	609
	T_2 (ms)	908	844	756	658	564
	η	1.93	1.79	1.57	1.33	1.11
	s_0	1	2.72	4.72	6.75	8.71
	s	1	2.58	4.14	5.38	6.25

Table 5c. Carboxylic ^{13}C ($>C{=}O$) $r_{C\cdots H} = 1.85$ Å; $\Delta\sigma = 250$ ppm. T_1, T_2, η, s, s_0, τ_c: see Table 5a

ν_0(MHz)		200	400	600	800	1000
$\left(\frac{B_0}{B_{200}}\right)^{3/2}$		1	2.83	5.20	8	11.18
$\tau_c = 5 \times 10^{-9}$	T_1(s)	0.61	0.61	0.60	0.60	0.60
	T_2(s)	0.22	0.08	0.04	0.02	0.01
	η	57×10^{-2}	1.410	6.210	3.410	2.210
	s_0	1	1.0	0.93	0.83	0.75
	s	1	1.0	1.88	0.79	0.71
$\tau_c = 5 \times 10^{-10}$	T_1(s)	1.35	0.58	0.31	0.21	0.15
	T_2(s)	1.2	0.47	0.24	0.14	0.09
	η	0.70	0.12	3.210	1.210	610
	s_0	1	1.75	2.24	2.62	2.93
	s	1	1.16	1.36	1.56	1.74
$\tau_c = 5 \times 10^{-11}$	T_1(s)	11.6	4.75	2.40	1.42	0.93
	T_2(s)	10.7	4.20	2.08	1.22	0.8
	η	1.03	0.41	0.20	0.11	0.07
	s_0	1	1.75	2.26	2.67	3.01
	s	1	1.22	1.34	1.47	1.59

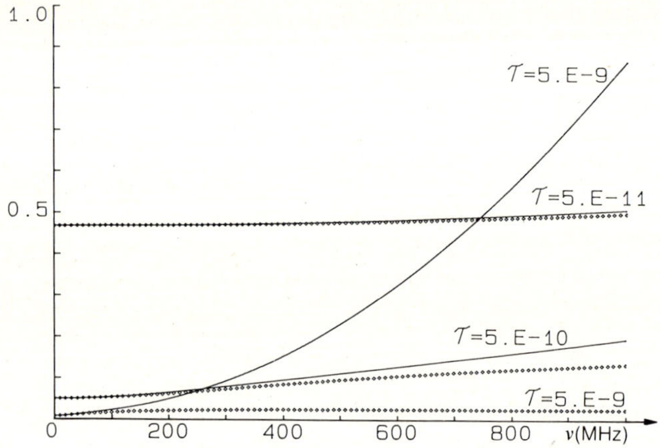

Fig. 7. Variation of the ^{13}C spin–lattice relaxation time T_1(s) (continuous line, Eqs. (30)–(27)) and of the ^{13}C spin–spin relaxation time T_2(s) (open squares, Eqs. (38)–(28)) for a C–C(H)(H) moiety. $r_{C\text{-}H} = 1.09$ Å, $\Delta\sigma = 25$ ppm. The correlation times are equal to 5×10^{-9} s, 5×10^{-10} s and 5×10^{-11} s respectively

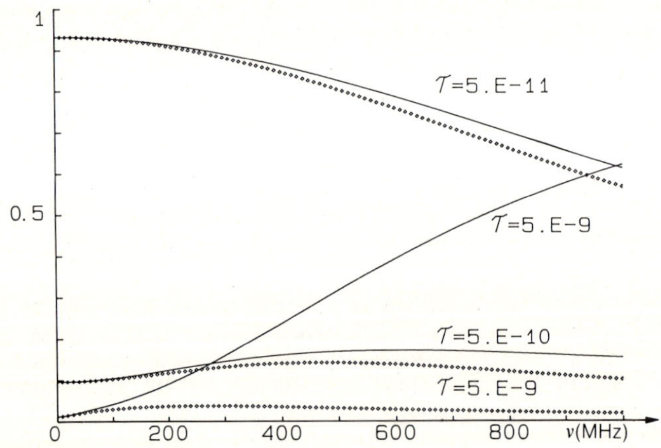

Fig. 8. Variation of the ^{13}C spin–lattice relaxation time T_1(s) (continuous lines Eqs. (30), (27)) and of the ^{13}C spin–spin relaxation time T_2(s) (open squares, Eqs. (38), (28)) for a >C=C<H moiety. $r_{C\text{-}H} = 1.09$ Å, $\Delta\sigma = 200$ ppm. The correlation times are equal to 5×10^{-9} s, 5×10^{-10} s, 5×10^{-11} s respectively

parameters with B_0: e.g. aliphatic carbons for correlation a time of 5×10^{-10} s without NOE.

The large increase of the carboxylic carbon relaxation rate R_2 for the small correlation times shows that the $^{13}C{=}O$ groups will be difficult to observe at very high fields. NMR spectroscopists are already facing this problem at

Table 6. Expression of the Log of the ^{13}C sensitivity (with (NOE)) with respect to the Log of the magnetic field Bo, Log s = n Log Bo + a for different values of the correlation time, between $\nu_0 = 400$ MHz and 1000 MHz

Correlation time (s)		5×10^{-9}	5×10^{-10}	5×10^{-11}
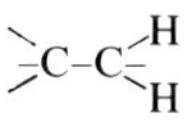	n	0.73	1.31	1.50
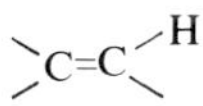	n	0.40	0.92	1.16

600 MHz. For the aliphatic carbons, which correspond to the type of carbon, biochemists are the most interested in, T_2 increases or remains constant in the considered range of correlation times.

5.3 *Nitrogen* ^{15}N

Data concerning ^{15}N magnetic shielding $\Delta\sigma$ are not too common in the literature and most of the data concern molecules of biological interest [56, 57, 58]. Moderate values of the order of 100–200 ppm seem to be the rule, except for triply bounded nitrogen. In Table 7 are given the evolution of the relaxation times and sensitivity as a function of the frequency, for slow and fast motions, and for typical N–H bond distance and magnetic shielding anisotropy. The

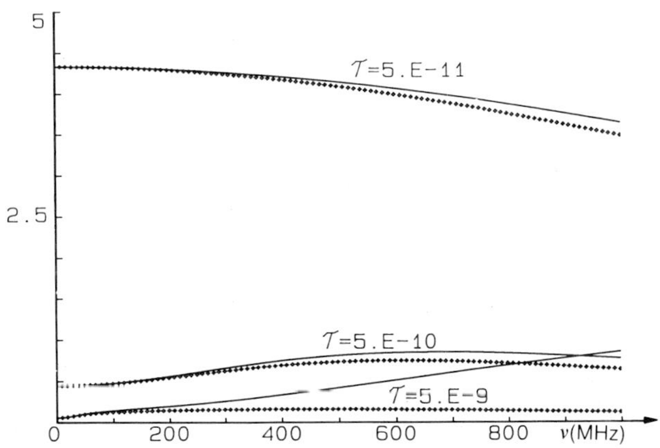

Fig. 9. Variation of ^{15}N spin–lattice relaxation time T_1(s) (continuous line, Eqs. (30), (27)) and of the ^{15}N spin–spin relaxation time T_2(s) (open squares, Eqs. (38), (28)) for a –NH moiety. $r_{N\text{-}H} = 1.04$ Å, $\Delta\sigma = 150$ ppm. The correlation times are equal to 5×10^{-9} s, 5×10^{-10} s, 5×10^{-11} s respectively

Table 7 ^{15}N. T_1(s) spin–lattice relaxation time; T_2(s) spin–spin relaxation time; η NOE factor (Eq. 33); s_0 sensitivity without NOE; s sensitivity with NOE. s_0 and s are normalized to 1 at 200 MHz. $-^{15}N\langle^{H}$; $r_{N-H} = 1.04$ Å; $\Delta\sigma = 150$ ppm; τ correlation time (s)

ν_0(MHz)		200	400	600	800	1000
$\left(\frac{B_0}{B_{200}}\right)^{3/2}$		1	2.83	5.20	8	11.18
$\tau_c = 5 \times 10^{-9}$	T_1(s)	0.18	0.32	0.50	0.68	0.84
	T_2(s)	0.13	0.15	0.14	0.13	0.11
	η	−0.66	−0.30	−0.21	−0.16	−0.13
	s_0	1	2.60	3.86	4.61	4.99
	s	1	5.37	9.04	11.46	12.90
$\tau_c = 5 \times 10^{-10}$	T_1(s)	0.52	0.68	0.78	0.79	0.74
	T_2(s)	0.52	0.65	0.71	0.69	0.62
	η	−4.60	−3.57	−2.42	−1.56	−1.00
	s_0	1	3.17	6.03	9.04	11.9
	s	1	2.26	2.39	1.40	−0.01
$\tau_c = 5 \times 10^{-11}$	T_1(s)	4.29	4.19	4.03	3.82	3.58
	T_2(s)	4.28	4.16	3.97	3.73	3.46
	η	−4.88	−4.72	−4.48	−4.17	−3.81
	s_0	1	2.78	4.99	7.44	9.99
	s	1	2.67	4.48	6.07	7.24

relaxation times T_1 and T_2 show only a small dependence with the frequency, with the exception of the lengthening of T_1 at high field for slow motions. The linewidth remains relatively small at high field, it is only for $\tau_c > 10^{-9}$ s that it exceeds 1 Hz for a ^{1}H observation frequency of 1 GHz. The negative sign of the ^{15}N magnetogyric ratio appears as a feature which deserves to be carefully examined at high field, since the NOE factor is itself negative and equal to $-9.87\ \sigma\ T_1$, (σ cross-relaxation term, T_1 spin lattice relaxation time (Eq. (33)). A well known consequence is a possible signal decrease or even disappearance (e.g. at 1 GHz for $\tau_c = 5 \times 10^{-10}$ s). Conversely, the combination of relaxation parameters induces a strong sensitivity gain with respect to $B_0^{3/2}$ around 500 MHz for $\tau_c = 5 \times 10^{-9}$ s. The sensitivity, without NOE is always increasing with the frequency. The sensitivity with NOE increases for slow and fast motions, but it shows a non monotonous variation for intermediate values of the correlation time. As an example, for $\tau_c = 5 \times 10^{-10}$ s, s exhibits a maximum for $\nu = 500$ MHz. The NOE enhancement factor and sensitivity dependence with the frequency are shown in Fig. 9.

5.4 Phosphorus (^{31}P)

From the NMR point of view the phosphorus compounds may be divided in two broad categories. On the one hand the non phosphate-like derivatives, including i.a. the typical organo-phosphorus compounds and the metal-phosphorus complexes; on the other hand, the compounds containing a phosphate group, including nearly all the molecules of biological and geochemical interest. The NMR of non biological compounds is relatively easy, because of the high ^{31}P receptivity R, as compared to other nuclei ($R(^{31}P)/R(^{1}H) = 0.066$) [59], and of the large range observed for the resonance frequency [60], [61], which is close to 600 ppm for the usual phosphorus chemical functionality. The ^{31}P NMR signals are most of the time readily obtained and are a clear signature of the chemical species under study. The advantage of going to high field thus concerns essentially the sensitivity, as the species under concern reorient sufficiently rapidly and have only one or a few ^{31}P spectral lines. Care must however sometimes be taken in tailoring the pulse experiment for relatively long T_1 and short T_2 (see Sect. 4).

Conversely, the case of the phosphate moieties is not favorable for NMR studies. Data concerning the magnetic screening constant and the components of the σ tensor may be found in Refs. [62–65]. The whole range of σ_{iso} is restricted to about 20 ppm, and this can make precise chemical identification difficult. A way to overcome this difficulty is to determine the full $\boldsymbol{\sigma}$ tensor or its principal values as a probe of the local symmetry. This approach is fruitful for biochemical and geochemical compounds. The chemical shift anisotropy $\Delta\sigma$ may range over 100 to 300 ppm. If a large $\Delta\sigma$ value may be turned into advantage for identification of species in oriented phase, it has more negative effects in high resolution NMR at high field. Phosphate moieties have several features which are not in favor of very high field NMR: (i) moderate to large $\Delta\sigma$ values; (ii) no effective dipolar relaxation owing to large non bonded P···H distances; (iii) often the PO_4 moiety belongs to a rigid part of the molecule, yielding large τ_c values. The variation of T_1 and T_2 as a function of the resonance frequency

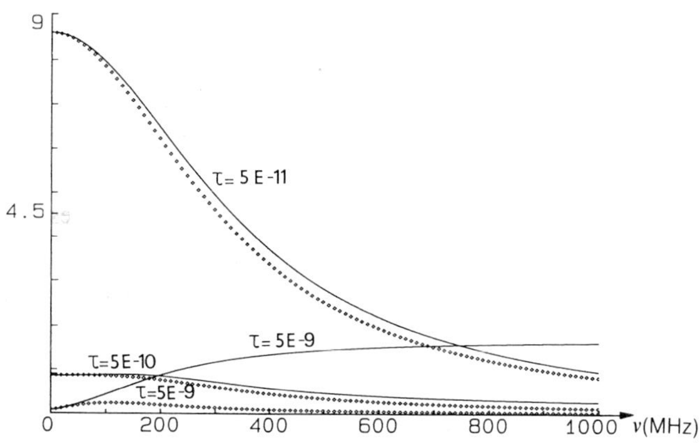

Fig. 10. Variation of ^{31}P spin–lattice relaxation time T_1(s) (continuous line, Eqs. (30), (28)) and of the ^{31}P spin–spin relaxation time T_2(s) (open squares, Eqs. (38), (28)) for a $-P\langle^{H}$ moiety. $r_{P-H} = 1.85$ Å, $\Delta\sigma = 150$ ppm

Table 8. ^{31}P T_1(s) spin–lattice relaxation time; T_2(s) spin–spin relaxation time; η NOE factor (Eq. 33); s_0 sensitivity without NOE, s with NOE, s_0 and s are normalized to 1 at 200 MHz. $-\mathrm{P}\!<^{\mathrm{H}}$; $r_{P\cdots H} = 1.85$ Å; $\Delta\sigma = 150$ ppm

ν_0(MHz)		200	400	600	800	1000
$\left(\frac{B_0}{B_{200}}\right)^{3/2}$		1	2.83	5.20	8	11.18
$\tau_c = 5 \times 10^{-9}$	T_1(s)	0.87	1.33	1.50	1.57	1.60
	T_2(s)	0.18	7.7×10^{-2}	3.9×10^{-2}	2.3×10^{-2}	1.510
	$\eta \times 10^2$	2.3	0.62	0.29	0.16	0.10
	s_0	1	0.99	0.86	0.77	0.69
	s	1	0.97	0.85	0.75	0.68
$\tau_c = 5 \times 10^{-10}$	T_1(s)	0.84	0.57	0.39	0.30	0.26
	T_2(s)	0.75	0.44	0.25	0.16	0.12
	η	0.60	0.13	3.4×10^{-2}	1.2×10^{-2}	0.5×10^{-2}
	s_0	1	2.1	2.8	3.3	3.6
	s	1	1.5	1.8	2.1	2.3
$\tau_c = 5 \times 10^{-11}$	T_1(s)	6.5	3.7	2.2	1.4	0.9
	T_2(s)	6.2	3.4	1.9	1.2	0.8
	η	0.92	0.52	0.29	0.17	0.11
	s_0	1	1.21	2.9	2.35	2.40
	s	1	1.6	1.9	2.1	2.3

ν are shown in Fig. 10, for a $\Delta\sigma$ value of 150 ppm. The spin–spin relaxation time T_2 is decreasing with ν, thus the linewidth increases. Although it does not increase as B_0^2, owing to the dipolar contribution, it reaches values which clearly indicate that for large correlation times, the ^{31}P nmr signal will become difficult to observe at high frequency. For fast motion, the line broadening may be partly compensated by the increase of other factors involved in the sensitivity (see Table 8). The conclusion of this section is that the ^{31}P nucleus is one for which very high field nmr is moderately interesting.

6 Conclusion

In this chapter we have presented a general discussion concerning the behaviour of the dynamic NMR parameters and of the sensitivity as a function of the magnetic field. The expressions of the relaxation times T_1 and T_2, of the cross relaxation terms parallel and perpendicular, namely σ and σ_ρ, and of the NOE factor η are given as a function of the resonance frequency.

The values of the above mentioned parameters have been calculated for the nuclei most encountered in molecules of biological interest (^{1}H, ^{13}C, ^{15}N, ^{31}P). The values chosen for the distances between the nucleus under study and the nuclei which cause the dipolar relaxation, and for the magnetic shielding tensor anisotropy correspond to realistic situations. A description of most of the models which have been proposed to describe the molecular motions has been presented. Nevertheless, for the numerical applications we have chosen the simplest form for the spectral density $\tilde{J}(\omega) = 2\tau_c/(1 + \omega^2\tau_c^2)$, with correlation time values τ_c equal to 5×10^{-9} s, 5×10^{-10} s and 5×10^{-11} s respectively. For the two most important nuclei, namely ^1H and aliphatic ^{13}C, T_1 and T_2 exhibit the well known increase with frequency. This has two major consequences concerning the sensitivity of the NMR experiments: on the one hand the natural linewidth decreases and on the other hand the waiting time between pulses in the multiscan mode must be increased in order to reach the magnetization recovery. For the olefinic ^{13}C, for ^{15}N and for ^{31}P, owing to their magnetic shielding anisotropy, the relaxation time T_2 first increases and then decreases when the applied magnetic field increases, with a maximum position which depends on the correlation time and on the magnetic shielding anisotropy (see Figs. 8–10). The changes are not that important from 200 MHz to 1 GHz for the aliphatic ^{13}C, but raise serious problems for observing ^{15}N and ^{31}P at very high field.

In every case (with the exception of ^{31}P with low correlation time values), the sensitivity with or without NOE, namely s or s_0, increases with the frequency.

Relaxation parameters are seen to play a major role on sensitivity, as soon as correlation times exceed 10^{-10} s; for smaller values, the sensitivity follows a normal dependence with respect to B_0($B_0^{3/2}$ which is considered here as a reference). Generally, one notices a loss with respect to $B_0^{3/2}$ although a significant increase can be observed in favorable instances (as an example, consider the aliphatic ^{13}C signals without NOE for $\tau_c = 5 \times 10^{-10}$ s). When CSA is important, severe sensitivity losses occur (as expected, e.g. for ^{31}P), even in the presence of dipolar contributions which could have been thought to somewhat counteract CSA effects (this is especially visible for an olefinic carbon and a correlation time of 5×10^{-9} s whose absolute sensitivity is not better, or even slightly worse, at 1000 MHz than at 600 MHz). There exists however noticeable exceptions concerning nuclei of negative gyromagnetic ratio observed with NOE, exemplified by ^{15}N, whose sensitivity drops to zero for $\tau_c = 5 \times 10^{-10}$ s but dramatically increases and largely exceeds $B_0^{3/2}$ for $\tau_c = 5 \times 10^{-9}$ s.

7 References

1. Bendall MR, Doddrell DM (1979) J Magn Reson 33: 659
2. Volino F (1977) In: Microscopic structure and dynamics of liquids. Nato Advanced Study Institute Series
3. Bloch F (1946) Phys Rev 70: 460
4. Solomon I (1955) Phys Rev 99: 559
5. Rose M (1976) Elementary theory of angular momentum. Wiley, New York
6. Abragam A (1961) Principles of nuclear magnetism. Oxford University Press
7. Redfield AG (1965) In: Waugh JS (ed) Advances in magnetic resonance, vol 1. Academic Press, New York, p 1
8. Szymanski S, Gryff-Keller AM, Binsch G (1986) J Magn Reson 68: 399

9. Spiess HW (1978) In: Diehl P, Fluck E, Kosfeld R (eds) NMR Basic principles and progress, vol 15. Springer, Berlin Heidelberg New York, p 59
10. Hubbard PS (1969) Phys Rev 180: 319
11. Vold RL, Vold RR (1978) In: Emsley JW, Feeney J, Sutcliffe LH (eds) Progress in NMR spectroscopy, vol 12, p 79. Pergamon, Oxford
12. Werbelow LG, Grant DM (1977) In: Waugh JS (ed) Advances in magnetic resonance, vol 4, p 189. Academic, New York
13. Hubbard PS (1970) J Chem Phys 52: 563
14. Haeberlen U (1976) In: Waugh JS (ed) Advances in magnetic resonance. Academic Press, New York, London (suppl 1) High resolution NMR in solids
15. Noggle DH, Schirmer RE (1971) The nuclear Overhauser effect. Academic Press, New York
16. Vold RR, Vold RL (1976) J Chem Phys 64: 1320
17. Mackor EL, Maclean C (1967) In: Emsley JW, Feeney J, Sutcliffe LH (eds) Progress in NMR spectroscopy, vol 30, p 129. Pergamon Press, Oxford
18. Goldman M (1984) J Magn Reson 60: 437
19. Gueron M, Leroy JL, Griffey RH (1983) J Am Chem Soc 105: 473
20. Farrar TC, Locker IC (1987) J Chem Phys 87: 3281
21. Nery H, Canet D (1981) J Magn Reson 42: 370
22. Werbelow LG, Marshall AG (1973) Chem Phys Lett 22: 568
23. Jones GP (1966) Phys Rev 148: 811
24. Bothner-by AA, Stephens RL, Lee J, Warren CD, Jeanloz RW (1985) J Am Chem Soc 106: 811
25. Bax A, Davis DG (1985) J Magn Reson 63: 207
26. Halle B, Wennerstrom H (1981) J Magn Reson 44: 89
27. Hubbard PS (1970) J Chem Phys 53: 985
28. Delville A, Detellier C, Laszlo P (1979) J Magn Reson 34: 301
29. Werbellow LG (1979) J Chem Phys 70: 5381
30. Werbellow LG, Marshall AG (1981) J Magn Reson 43: 443
31. Huntress WT Jr (1970) In: Waugh JS (ed) Advances in magnetic resonance, vol 4. Academic Press, New York. Steele WA (1976) In: Prigogine I, Rice AR (eds) Advances in chemical physics, vol xxxiv, p 1. John Wiley and Sons, New York
32. Hertz HG (1985) In: Emsley JW, Feeney J, Sutcliffe LH (eds) Progress in NMR spectroscopy, vol 16, p 115. Pergamon Press, Oxford
33. Egelstaff PA (1967) An introduction to the liquid state. Academic Press, London
34. Gordon RG (1966) J Chem Phys 44: 1840
35. McClung RED (1969) J Chem Phys 55: 3459
36. McClung RED (1972) J Chem Phys 57: 5478
37. Bull TE, Egan W (1984) J Chem Phys 81: 3181
38. Fixman M, Rider K (1969) J Chem Phys 51: 2425
39. Hubbard PS (1972) Phys Rev A6: 2421
40. McClung RED (1980) J Chem Phys 73: 2435
41. McClung RED (1980) J Chem Phys 75: 5503; GT Evans (1977) J Chem Phys 67: 2913
42. Lee DH, McClung RED (1987) Chem Phys 112: 1
43. Bull TE (1988) Chem Phys 121: 1
44. Perrin F (1934) J Phys Radium V: 33
45. Perrin F (1936) J Phys Radium VII: 1
46. Woessener DE (1961) J Chem Phys 36: 1
47. Huntress WT Jr (1968) J Chem Phys 48: 3524
48. Woessner DE (1962) J Chem Phys 37: 647
49. Wennerstrom H, Lindman B, Soderman O, Drakenberg T, Rosenholm JB (1979) J Amer Chem Soc 101: 6860; Lippari G, Szabo A (1982) J Amer Chem Soc 104: 5146
50. Ernst RR (1966) In: Waugh JS (ed) Advances in magnetic resonance, vol 2, pp 1–131. Academic Press, New York
51. Hoult DI, Richards RE (1976) J Magn Reson 24: 71
52. Mehring M (1976) In: Diehl P, Fluck E, Kosfeld R (eds) NMR basic principles and progress, vol 11. Spinger-Verlag, Berlin Heidelberg New York
53. Ernst RR, Bodenhausen G, Wokaun A (1987) Principles of nuclear magnetic resonance in one and two dimensions. Clarendon Press, Oxford
54. Ducan TM (1987) J Phys Chem Ref Data 6: 125
55. Facelli JC, Grant DM (1989) In: Eliel EL, Wilen SH (eds) Topics in stereochemistry, vol 19, pp 1–61. An Interscience publication. J Wiley and Sons, New York

56. Harbison G, Herzfeld J, Griffin RG (1981) J Amer Chem Soc 103: 4752
57. Harbison G, Jelinski LW, Stark RE, Torchia DA, Herzfeld J, Griffin RG (1984) J Magn Reson 60: 79
58. Valentine KG, Rockwell AL, Gierash LM, Opella SJ (1987) J Magn Reson 73: 519
59. Harris RK (1978) In: Harris RK, Mann BE (eds) NMR and the periodic table, Academic Press, London
60. Crutchfield MM, Dungan CH, Letcher JH, Van Wazer JR (1967) In: Grayson M, Griffiths EJ (eds) Topics in phosphorus chemistry, vol 5, Chap. 4, Wiley-Interscience. New York
61. Tebby JC (1987) In: Verkade JG, Quin LD (eds) Phosphorus-31 NMR spectroscopy in stereochemical analysis, Chap 1. VCH Publishers, Deerfield Beach
62. Robert JB, Wiesenfeld L (1982) Phys Rep 86: 363
63. Robert JB, Wiesenfeld L (1987) In: Verkade JC, Quin LD (eds) Phosphorus-31 spectroscopy in stereochemical analysis. Chap 4, VCH Publishers Deerfeld, Beach
64. Gorenstein DG (1984) Phosphorus-31 NMR. Academic Press, San Francisco
65. Sun U, Klein MP (1989) J Amer Chem Soc 111: 5119

Structural Studies of Biomolecules at High Field

Dominique Marion

Centre de Biophysique Moléculaire, Centre National de la Recherche Scientifique, lA, Av. de la Recherche Scientifique, 45071 Orléans Cedex 2, France and Laboratory of Chemical Physics, National Institute of Diabetes and, Digestive and Kidney Diseases, National Institutes of Health, Bethesda MD 20892, U.S.A.

Table of Contents

Present address: Laboratoire de Biologie Structurale Centre d'Etudes Nucléaires Bat. C5 B.P. 85X 38041 Grenoble Cedex France

NMR Basic Principles and Progress, Vol. 25

1 The Various NMR Approaches for Biomolecules

In comparison with other spectroscopic techniques used in biophysical research (such as absorption, circular dichroism, infrared, raman, fluorescence or EPR...), NMR investigation of biomolecules can be divided into two classes, one that could be called the *conventional approach*; the second the *structural approach.*

In the first class, NMR is used just as any other spectroscopic technique in order to monitor a physical or chemical modification in the sample (change of temperature, addition of an interacting drug...). Here, most efforts are devoted to careful preparation of the sample; the NMR methods then required are rather limited compared to the skills of the investigator during the previous biochemical or chemical steps. In the second class, the decisive advantage of NMR over other techniques relies on the numerous sophisticated experiments which can be performed on the same sample, offering a great wealth of information on the conformation and dynamics of the macromolecule. Here, the quality of the results is determined by both the NMR methods and the purification or synthesis of the compounds. Historically the former class of investigation preceded the latter until recent developments of both high field magnets and two-dimensional (2D) NMR spectroscopy allowed the in-depth study of macromolecules.

The largest part of this review will be devoted to the structural approach on the following grounds. Because this book deals with high field magnets, it should be pointed out that the maximal benefit reaped from them has been structural studies of biopolymers by 2D-NMR. This choice is in no way based on the biological significance of the derived results. Indeed, some biologists could be more interested in evidence—even if only qualitative—of an interaction between two biomolecules than in the complete resolution of a protein structure.

In the conventional approach described above, NMR has been successfully applied to the detection of a variety of *conformational changes* or *intermolecular interactions*. For this purpose, the main advantage of using a high field magnet is the gain in sensitivity. As a matter of fact, NMR spectra always have to be recorded at a concentration much higher than required by other spectroscopic methods, because of the intrinsically weak sensitivity of NMR (only one nucleus in 10^5 present in the sample is detected!). Thus the biological relevance of the NMR investigation may be biased by possible intermolecular interactions, which are likely to occur in a different manner in biological conditions where the concentration of each type of molecules is likely to be weaker (10^{-6}–10^{-5} M) but where numerous other molecular species are present.

2 Interactions of Biomolecules Studied by NMR

2.1 Ion-Biomolecule Interactions

Illustrations of this type of problems are found in numerous investigations dealing with the interactions of biomolecules with cations [1, 2]. In the case of carriers or ionophores, for instance, these studies are run in parallel with investigations of membrane permeability modifications induced by these compounds at very low concentration (10^{-6} M or less). The interaction of valinomycin (a macrocyclic antibiotic) with cations has been investigated [3, 4] by NMR and CD at non biological concentrations: 1 mM for CD, 5 mM for ^{1}H NMR and 25 mM for ^{13}C NMR. The experimental conditions leave a lingering doubt as to whether the peptide-ion-peptide sandwich observed for Ca^{2+} [4] can be found in membranes or whether this originates only from the high concentration of both salts and antibiotic. Unfortunately, this problem will not be settled in a definite manner even if technology provides a two-fold increase in the magnetic field in the next ten years.

NMR methods are also unique in this field because the same phenomenon can be studied using both the ^{1}H spectrum of the biomolecule and that of the ion, provided that its abundance, spin number and (if applicable) quadrupolar moment are suitable for observation [5]. Variations in chemical shift or in relaxation times of several nuclei such as ^{7}Li [6], ^{113}Cd [7], ^{23}Na or ^{205}Tl [8] have provided insight into the complexation of biomolecules. For appropriate binding constants and binding rates, spectra can be obtained in the presence of an excess of ions, which may avoid self-association of the biomacromolecule. However, due to the low sensitivity of most heteronuclei (for lack of J-coupling, no polarization transfer can be achieved), the use of high-field magnets is mandatory.

2.2 Drug-DNA Interactions

Classical NMR spectroscopy has also been used extensively in attempts to characterize the binding site and thus the mode of action of several DNA-interacting drugs (actinomycin D [9, 10], acridine derivatives [11], netropsin [12] and distamycin A[13]). Whatever the molecular binding mechanism is (intercalation between bases or interaction in the grooves [14]), the chemical shifts of DNA base protons is an easy and sensitive probe, owing to their sensitivity to ring current shift (see below). During the late 1970s and early 1980s, dinucleotides were used as models [9, 1, 15] for investigating these interactions. The choice of such short fragments was dictated by their avilability and low cost, and their nearly resolved NMR spectrum. However, the dinucleotide systems investigated showed fast kinetics (i.e. the observed spectrum was an average of that of the free and complexed species [11]) and may not be truly representative of the slowly associating/dissociating binding mode seen in biological experiments. Another problem is that the geometries of the complexes (mainly based only on ring current shift analysis) were tentative. Compared to larger fragments, dinucleotides lack constraints on both sides of the interaction sites. In this research area, high fields, in combination with 2D NMR and automated synthesis of oligonucleotides, has made a more realistic approach possible using larger pieces of DNA (8 to 12 base pairs) [13, 16, 17].

2.3 Lipid Interactions in Membranes

NMR has provided a description of membrane lipid dynamics over a large time scale range. The biological membrane is a dynamic assembly of many distinct components (phospholipids, but also proteins, sterols, oligosaccharides...) with each component capable of influencing the others. Overall membrane structure, composition and phase equilibria as well as local properties detectable by NMR (*conformation, molecular order and dynamics*), all contribute to membrane function [18]. In either pure phospholipid systems or natural membranes, these local features have been studied by ^{1}H, ^{2}H, ^{13}C, ^{14}N, ^{31}P and ^{19}F (substituted for ^{1}H) NMR. Several reviews dealing with these topics have been published [19]. While some probes are naturally present in lipids, labelling, preferably at specific positions, is required in other cases (^{2}H). The induced local perturbation is expected to be negligible compared with that caused by the bulky probes necessary in EPR or fluorescence.

The lipid molecules undergo anisotropic rotational diffusion about their long axis and perpendicular to that axis, as well as lateral diffusion within the bilayer. Moreover, the two different moieties, the *head group* and the *hydrocarbon chain*, both have a high degree of motional freedom. ^{31}P is a suitable probe for the motion of the head group [20], and ^{13}C and ^{2}H for that of the hydrocarbon chain [21]. As is the case for other spectroscopic techniques, NMR has its own characteristic time scale, the lower limit of which is determined by the overall width of the spectrum and the upper limit by the width of each spectral component. However, the scale varies from one nucleus to another (for example the scale of ^{13}C or ^{31}P is larger than that of ^{2}H because their spectra are narrower). Depending on the nature of the motion, the shapes of NMR spectra of lipids in membranes provide insight into motions on a time scale ranging from $1\,\mu s$ to 10 ms. For the same nucleus, different relaxation times (T_1, $T_{1\rho}$ (*spin-locked relaxation time*) or T_2) are sensitive to different orders of magnitude of motional rates: T_1 reflects motion at the Larmor frequencies (ω_0 and $2\omega_0$), i.e. at 10^{-9} s, while $T_{1\rho}$ at motion at 10^{-4}–10^{-6} s, depending on the r.f. field strength. Thus, apart from an increase in sensitivity, no substantial advantage at very high field is expected, because the observation window of NMR—*with regard to the motional time scale*—will not be extended by an order of magnitude towards faster motions.

However, ^{13}C, ^{31}P, and ^{2}H have slightly different NMR properties, and thus their behavior at higher fields should be questioned. For ^{13}C bound to protons, relaxation is dominated by the dipole–dipole intercation with the attached protons and thus ^{13}C is observable at higher field without any change in the experimental conditions. ^{31}P and ^{2}H exhibit very large spectral widths, originating from different NMR features: The ^{31}P chemical shift of a rigid phosphodiester, such as found in lipids, depends strongly on the orientation of this group with respect to the magnetic field of the spectrometer (*chemical shift anisotropy*). Depending on the type of membrane phase and its motional characteristics, a powder spectrum with a width of 50 to 100 ppm is observed [22]. Since ^{2}H is a quadrupolar nucleus ($I = 1$), a residual *quadrupolar splitting* of several kHz is observed for partially oriented membranes. A deuterium order parameter, which characterizes the average orientation and the fluctuation of the C-^{2}H bond, is derived from this splitting.

In both cases, one is faced to the problem of exciting and detecting large spectral widths and inappropriate techniques may lead to spectra suffering from distortion and thus to erroneous conclusions. These problems, experienced in both ^{2}H and ^{31}P NMR, have been solved by echo experiments (quadrupolar echo for ^{2}H [23] and Hahn echo [24] for ^{31}P). Let me point out that the ^{31}P chemical shift is obviously proportional to the magnetic field, whereas the ^{2}H quadrupolar splitting is constant. As a result, the problems raised above will evolve differently at higher magnetic fields.

Finally, let it be remarked that recently, certain lipid membranes have been shown to align spontaneously in the field of NMR magnets of 7 T [25] and 9.3 T [26]. The driving force is the negative diamagnetic anisotropy of the fatty acyl chains, leading to an orientation of the membrane plane parallel to the magnetic field. (See Chap. 3 of this volume)

3 The Conformation of Biopolymers

In addition to ions and water, two different types of chemical substances are encountered in cells: small molecules and biopolymers. The former ones consist of fewer than 50 atoms and each substance has it own characteristic structure. Biopolymers are composed of subunits (called monomers or residues) linked in chains by covalent bonds. The three main types of biopolymers are nucleic acids, proteins and polysaccharides.

Nucleic acids [27] preserve and transmit genetic information. *Proteins* [28] are the working molecules of the cells (they catalyze chemical reactions, provide structural rigidity, regulate the concentrations of needed metabolites, recognize other biomolecules, control the functioning of genes...). *Polysaccharides or carbohydrates* [29, 30] play important roles as energy-storage molecules within the cells, as recognition sites or markers and as structural components when located on the outer surface of cells or between cells.

It is now well-established that the biological activity of a macromolecule is determined by its conformation and its ability to interact with its surroundings. Although both conformation and dynamics of macromolecules are primarily related to their covalent structure, most of them are able to adopt different conformations depending on the experimental conditions (temperature, interaction with the solvent or other molecules...). I will now discuss conformational questions of the three major classes of biopolymers and the potential of NMR for solving them.

3.1 Proteins

Amino acids, the building blocks of proteins, differ only in their side-chains. All 20 amino acids encountered in eukariotic cells are constructed according to a basic design: a central asymmetric carbon (α-carbon) is bonded to an amino group, to a carboxyl group, and to the variable side chain. The amino acids can be classified according to their electric charge when they are ionized (positive or negative) and according to their affinity for water (polar and non-polar groups). The peptide bond that covalently connects two amino acids in a protein exhibits a partial double bond character. Therefore, the six atoms of the peptidic group lie in the same plane ($\omega = 180°$). However, adjacent peptide groups are not coplanar due to rotation about the $C-C^{\alpha}$ and $N-C^{\alpha}$ bond. The backbone conformation of a protein is thus characterized by two dihedral angles per residue called ϕ and ψ [31].

Protein structure can be described at four different levels [32]. Primary structure refers to the linear arrangement of the residues along the polypeptide chain (the sequence of the protein). Secondary structure pertains to the folding of parts of the chains into regular structures such as α-helices (see Fig. 1) or β-sheets. Tertiary structure includes the folding of regions, which connect the helices and sheets, and thus concerns the location of these stereoregular structures with respect to one another. Finally, quaternary structure refers to the noncovalent interactions between individual proteins in a complex.

Whereas NMR is able to some extent to provide information on the sequence of relatively small proteins (for example determining the number of side-chain amide group), this is definitely not the most efficient technique for sequencing larger proteins (see [33] for sequencing techniques). However, because NMR easily reveals the occurrence of impurities (if they are at least several % of the total concentration), it can be used for monitoring sample purification by HPLC, as recently reported on a cardiotoxin from cobra venom [34]. NMR is more powerful in the determination of secondary structure and then, with the help of structure refinement programs, of tertiary structure. NMR yields more details of secondary structure in solution than any presently available technique (infrared, Raman or circular dichroism spectroscopy...). Furthermore, the structural

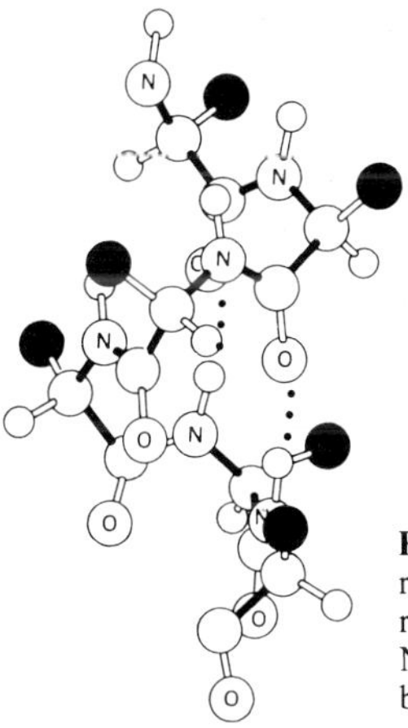

Fig. 1. A ball and stick model of an α-helix. the side-chains are represented by single black balls. There are about 3.6 amino acid residues per turn of the helix. An α-helical structure can be evidenced by NMR by small $^{3}J_{NH-C^{\alpha}H}$ (4 Hz) and the short internuclear distance between next-neighbor NH (2.8 Å)

elements identified can be located precisely within the sequence [35]. Let me finally remark, that the determination of quaternary structures as observed in larger proteins, such as multienzyme complexes, is far beyond the present potential of highfield NMR.

3.2 Nucleic Acids

There are two closely related information storing molecules in cells: deoxyribonucleic acids (DNA) and ribonucleic acids (RNA). Like proteins, DNA and RNA are linear biopolymers; however, the number of monomers in nucleic acids is generally much greater than in a protein, ranging from 80 to millions. Therefore, relative to intact DNA molecules, only very short fragments can be studied by NMR such as 20 base-pair DNA or transfer RNA. Whether or not this small fragment has functional significance may depend on the sequence.

Both DNA and RNA consist of only four different monomers called nucleotides. A nucleotide has three parts: a phosphate group, a pentose (a five-carbon sugar), and an organic base. The bases, either a purine or a pyrimidine, are adenine (A), guanine (G), cytosine (C) and thymine (T), the latter being replaced by uracyl (U) in RNA. The native state of DNA is a right-handed double helice of two antiparallel chains (A or B DNA) [36] that have complementary sequences of nucleotides (a purine always pairs with a pyrimidine). Certain short DNA polymers can adopt an alternative left-handed conformation called Z-DNA [37, 38].

The conformation of nucleic acids can be described in two complementary ways, the first by means of local torsion angles (which make the NMR interpretation easier) and the second by means of helical parameters. The local conformation of a nucleotide (strongly related to that of its complementary one)

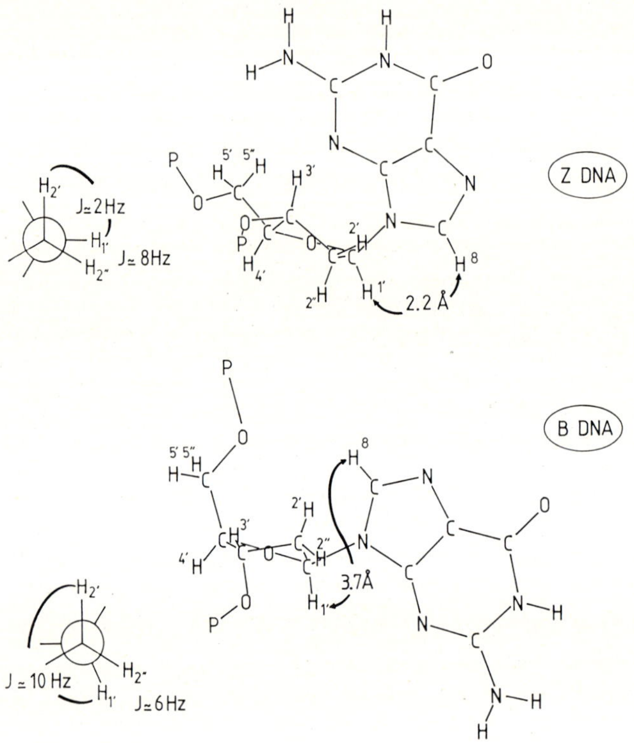

Fig. 2. Conformation of DNA and NMR spectral parameters. The conformation of guanosine is depicted as shown in Z–DNA (*C3′ endo* pucker for the ribose and *syn* orientation about the glycosidic bound) and in B–DNA (*C2′ endo* pucker and *anti* orientation). The sugar pucker can be determined by the vicinal coupling constant between the H1′ and the H2′ and H2″ (see simulation in Fig. 17) and the orientation about the glycosidic bond by the N.O.E. intensity between H1′ and the base proton H8

can be characterized by the following features (see Fig. 2): 1) the *sugar pucker* [39] (the five membered furanose is generally nonplanar), 2) the *orientation about the glycosyl bond* (i.e. the base with respect to the sugar) and 3) the conformation of the *phosphodiester backbone* (the C4′–C5′ bond). On the other hand, nucleotides also display order which can be expressed in terms of helical parameters: the pitch of the helix, the twist and roll of a base pair, and the propeller twist within each pair (see, for example [40]).

As a matter of fact, the set of possible conformations in DNA duplexes is much more limited than in the case of proteins. Conformational studies of oligonucleotides can be classified into two approaches: the sequence dependence of the local conformation of duplexes, and the study of either modified sequences (such as mismatches) or drug-DNA complexes. In contrast, possibilities in investigating transfer RNA are more varied because each nucleotide exhibits a different conformation.

3.3 Carbohydrates

As their name indicates, carbohydrates are constructed of carbon, hydrogen and oxygen [29]. The monosaccharides, often called simple sugars, have the structure $(CH_2O)_n$ with n = 3, 4, 5... and all contain an aldehyde or keto group. The most common monosaccharides are pentoses (n = 5, present in DNA and RNA) and hexoses (n = 6). A hexose such as glucose can exist in three different forms: a linear structure, a six-membered ring (pyranose form, the most abundant) and a five-membered ring (furanose form). The chair is the most stable conformation of the pyranose ring [30]. Due to the asymmetry of the 1 carbon, a pyranose can crystallize, depending on conditions, in two forms with different melting points and optical rotation: the α and β D-pyranose.

Disaccharides consist of two monosaccharides linked together by a glycosidic bond (see Fig. 3). In fact, there is a large number of possible linkages, since the 1 carbon of one sugar can react with any of the hydroxyl groups of the second one. Compared to proteins and nucleic acids, the differences in the chemical structure among the various common oligosaccharides are slight because most of the carbons bears a hydrogen and an OH group, and consequently induce only small chemical shift variations. Let me point out that the flexibility of oligosaccharides, which leads to conformational averaging, is another major limitation for structural studies. Resonance assignment of proton spectra of oligosaccharides is still difficult [41] and thus NMR has been used mainly for probing the primary structure rather than for extensive conformational studies. This research area is expected to reap maximum benefit from higher fields, which will enable the study of larger oligosaccharides (more than 5 or 7 sugars).

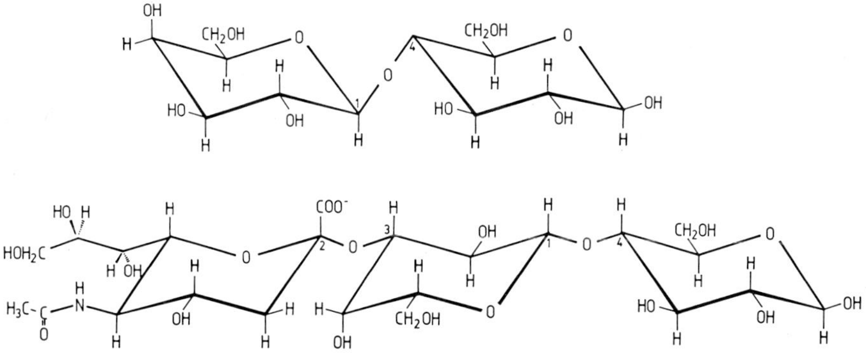

Fig. 3. Examples of oligosaccharides. On the top, lactose, the major sugar in milk, exhibits a β(1–4) linkage. On the bottom, the structure of α-Neu5Ac-(2–3) βGal-(1–4)Glc, a trisaccharide which has been assigned by Lerner and Bax [162] by means of heteronuclear correlation (see Fig. 21)

4 Useful Nuclei in 2D NMR of Biomolecules

Four different nuclei encountered in the aforementioned biopolymers are practical candidates for structural studies: ^{1}H primarily, but also ^{31}P, ^{13}C, and ^{15}N [42]. Their natural abundance, sensitivity. chemical shift dispersion, relaxation mechanisms and magnitude of scalar coupling differ widely. The use of a particular nucleus will be addressed with regard to its advantages and drawbacks.

The *natural abundance* of a nucleus is a key feature of its sensitivity [5]. Because ^{1}H and ^{31}P are very abundant in contrast with ^{13}C (1%) and ^{15}N (0.36%), the required concentrations of biological material will be highly dependent upon the observed nucleus (solubility limitations may be encountered). For ^{13}C or ^{15}N, isotopic enrichment is required for experiments at low concentration. By the way, provided that the chemical or biochemical technology is actually available, site-specific labelling may be of great help in simplifying the spectrum or analyzing biosynthetical pathways. It will be shown later see Fig. 22 that the ^{1}H spectrum of a large molecule can be indirectly simplified in correlation experiments with an heteronucleus nucleus. Because of the low abundance of ^{13}C or ^{15}N, no heteronuclear coupling are observed in a conventional ^{1}H spectrum. On the other hand, this asymmetry in natural abundance implies a more important one in the design of heteronuclear correlation experiments: only one ^{1}H over 100 is coupled to a ^{13}C, but virtually all non-quaternary ^{13}C are coupled with a proton.

As determined from a single pulse experiment, the intrinsic *NMR sensitivity* of a spin-1/2 nucleus is directly proportional to the third power of its magnetogyric ratio γ. The balance is here again in favor of ^{1}H relative to the other nuclei. The goal of many pulse experiments is signal enhancement of a low γ spin by means of magnetization or polarization transfer from ^{1}H. The efficiency of these techniques depends on the magnitude of the spin interactions and thus on the chemical structure of the macromolecule (a C–H can be more easily detected than a C=O), and it may be hard to define the actual sensitivity of a spin in general terms. Because of this large sensitivity difference, attention should be paid to the correct design of any ^{1}H–X correlation experiment, in order to ensure the highest signal-to-noise ratio by directly detecting the ^{1}H signal and not that of X [43–46]. This may result in a dramatic sensitivity gain by a factor $(\gamma_H/\gamma_X)^{5/2}$ ranging from 30 in the case of ^{13}C to 300 in the case of ^{15}N [47].

The *chemical shift dispersion* of an isotope is another severe limitation, which becomes more pronounced in larger biopolymers. ^{1}H has a very limited overall range of chemical shifts (10 to 15 ppm) compared to ^{15}N (1000 ppm), to ^{13}C (300 ppm) or to ^{31}P (700 ppm). However, as a result of the polymeric nature of biomacromolecules, no large variation of the chemical structure occurs along the molecule, so the actual chemical shift range for all nuclei is usually much narrower than the theoretical one. For example, the ^{1}H resonances of protein amide groups may fall within 2 or 3 ppm as shown in Fig. 4 but the corresponding ^{15}N resonance will spread over 25 ppm [48]. The larger range for nuclei other

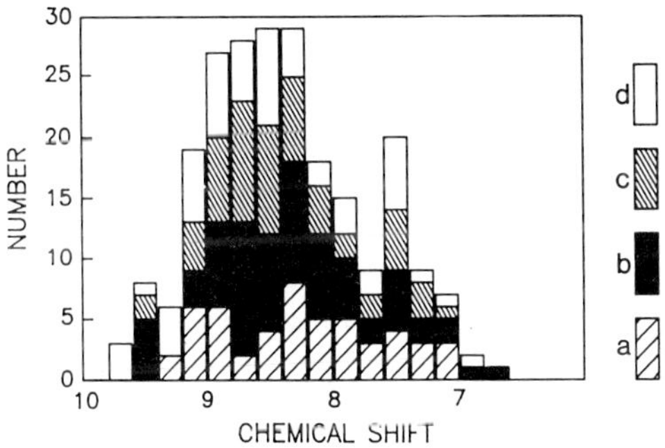

Fig. 4. Chemical shift range of the NH in proteins: *a*) Cardiotoxin CTXIIb form venom of the cobra *Naja mossambica mossambica* (115) *b*) Metallothionein-2 of rat-liver (116) *c*) Tendamistat, an α-amylase inhibitor from *Streptomyces tendae* (117) *d*) Ribosomal protein E-L30 of *Escherichia coli* (118)

than ^{1}H is of great help for assigning the overcrowded ^{1}H spectrum through heteronuclear correlation (see below).

The dominant *relaxation mechanism* (as well as its relative efficiency) may enhance or limit the use of a given nucleus for conformational analysis. For successful conformational analysis based on relaxation data, two conditions should be fulfilled. First, the relaxation should be mainly dipolar (other mechanisms are not distance-dependent) and second, the distance involved in the pathway should be a conformational probe. These requirements are only met for ^{1}H, provided that efforts has been made to remove either intrinsic or extrinsic paramagnetic species (dipolar relaxation with an unpaired electron). Let me remark that the rare ^{13}C and ^{15}N spin have no impact on ^{1}H relaxation.

As far as ^{13}C, ^{15}N or ^{31}P are concerned, distinction should be made between protonated nuclei whose relaxation is mainly dipolar, and non-protonated ones whose relaxation is dominated by *chemical shift anisotropy (CSA)*, the rate of which is proportional to $\Delta\sigma^2 B_0^2$ becoming increasingly dominant at high fields. ^{31}P never bears a proton in nucleotides and thus the CSA contribution [49] may amount to roughly 70% (for poly(I) at 6.4 T [50]) and more than 90% in larger fragments such as $^{t}RNA^{Phe}$ at 8.4 T [51]. The T_1 relaxation of protonated ^{13}C and ^{15}N does not provide conformational information because of the fixed C–H or N–H bond lengths, but is a suitable probe of the motional properties of biomolecules. The anisotropy of molecular tumbling or the internal flexibility can be investigated (refer to ref. [52–54]) for theoretical aspects and [55–56] for applications).

With regard to the relaxation times, the *repetition rate* in 2D experiments is a major concern because a compromise has to be found for avoiding artifacts while minimizing total acquisition times. If the relaxation delay is too long, experimental time is wasted; but if it is too short, the sensitivity and the quality of the data may also drop. A fairly relaxed longitudinal magnetization (delay $>T_1$ or $2T_1$) is a good starting point of many 1D or 2D experiments. For very large macromolecules, where $T_2 \ll T_1$, the worst possible situation is encountered: low sensitivity because of the long T_1 and poor resolution because of the short T_2. The design of some heteronuclear experiments dealing with a slowly relaxing C=O faces a similar problem.

The *scalar coupling constants* is the final NMR parameter I will discuss with respect to the two goals of 2D NMR of biomolecules: resonance assignment and conformational analysis. It is convenient to discriminate between the large $^1J_{XH}$ (typically 90 Hz for nitrogen and 120–140 Hz for carbon) and the smaller long-range coupling $^2J_{XH}$ and $^3J_{XH}$. Because of their large magnitude, the one-bond coupling constants easily provide connectivities between ^1H and ^{13}C or ^{15}N ones. Unfortunately, no structural data is carried by the ^1J. For detection and measurement of homo- or heteronuclear ^2J or ^3J, one has to cope with the line width of the resonances, which can be of the same order of magnitude. A J-coupling, which can be easily measured on a small molecule, may be undetectable in a large protein. Running spectra at higher field does not affect the scalar coupling and will have little effect on the line width: the conflict seems thus without resolution. ^2J only provides assignment information, but the *vicinal coupling* ^3J depends upon the involved torsion angle via a Karplus-like relationship and has been used for a long time in conformational analysis of peptides [57], proteins, nucleotides and oligosaccharides.

5 Assignment Strategies for Biopolymers

Early investigators of biomolecules were faced with two major problems: the inability to resolve individual lines in the ^1H spectrum of biopolymers, (see for example the 40 MHz spectrum of ribonuclease [58]) and the lack of efficient and reliable methods for assignment of the resolved lines. The availability of high field spectrometers has improved spectral resolution, but the second stumbling block would still remain without the advent of 2D NMR spectroscopy. In fact, the measurement of any spectral parameter is utterly devoid of interest as long as this piece of information cannot be converted into structural data through resonance assignment.

Many different procedures, somewhat reliable, have been tried in the past for assigning resonances of biopolymers. One such method is the comparison of the ^1H spectrum of a n-residue peptide with that of the previously assigned (n – 1)-residue analog. In cases where the structure has been solved previously from X-ray data, ring current shift calculations, or the introduction of a paramagnetic shift or relaxation reagent can provide assignment, albeit usually for only a small part of the molecule. These strategies, which rely on assumptions that there

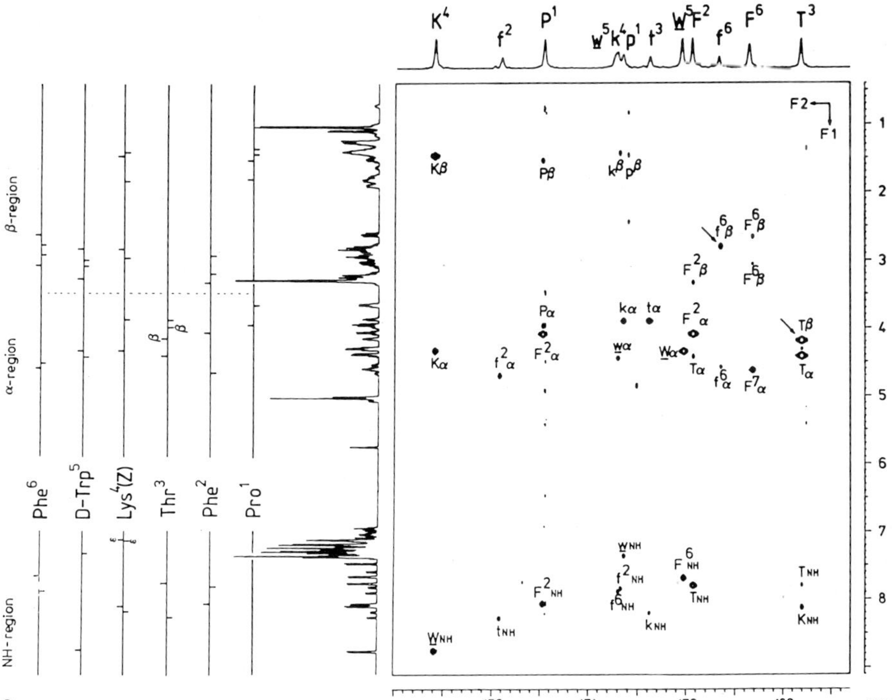

a

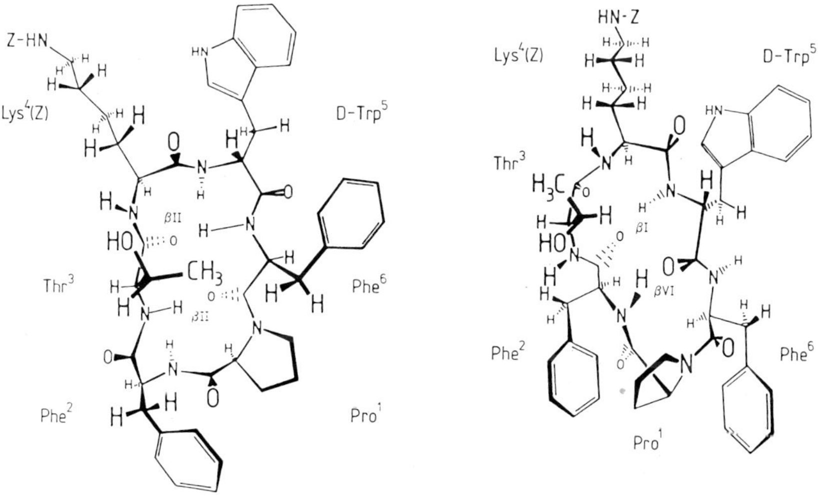

b

Fig. 5 a, b. 500 MHz $^{1}H_{4}-^{13}C$ COLOC spectrum of cyclo (Pro^1 -Phe^2 -Thr^3 -Lys^4(Z)-DTrp^5 -Phe^6) in DMSO. Note that each C=O resonance exhibits connectivities with protons of its own residue as well as with the NH of the following one. This peptide adopts two conformations which slowly interconvert. The predominant conformation (structure on the left in **b**) corresponds to the cross-peaks in **a**) labelled with upper case. Reprinted with permission from Kessler et al. [61]. Copyright American Chemical Society (1987)

is no difference between the crystal and the solution structure, lead to circular arguments with regard to the conclusions from the NMR data.

Inspection of the covalent structure of the three major types of biopolymers reveals a common feature. One or more ^{1}H *spin systems* (i.e. a group of spins connected by scalar coupling) are found within each monomer, but no ^{1}H spin system extends beyond a monomer. Experiments capable of delineating J-coupling provide only spin–spin connectivities within a amino acid, a deoxyribose ring of an oligonucleotide or a sugar, but the link between the monomers has to be bridged by other means: either a heteronuclear coupling constant or a ^{1}H–^{1}H nuclear Overhauser effect (N.O.E).

As far as peptides (and eventually proteins) are concerned, ^{13}C as well as ^{15}N coupling constants with ^{1}H can be used for establishing a link between next-neighbor residues. Each carbonyl ^{13}C exhibits two long-range $^2J_{CH}$, one with the following NH and another with its own C$^\alpha$H [57]. Similarly, each amide ^{15}N is coupled—in addition to the $^1J_{NH}$—with the C$^\alpha$H of the preceding residue [59]. Note, by the way, that this $^3J_{NH}$ provides the only estimate available for the ψ angle in a peptide. As reported in the previous section, the low abundance of these nuclei and the magnitude of the J-coupling presently prevent the use of this technique in proteins with broadened resonances. Nonetheless, in small peptides, this is an elegant alternative to the N.O.E., which can be vanishingly small in medium sized molecules [60]. Kessler and coworkers [61–63] have assigned and investigated the conformations of several peptides and glycopeptides by ^{1}H–^{13}C long range correlation (see Fig. 5).

In a similar way—but with a success guaranteed by the natural abundance of ^{31}P—internucleotide connectivities can be revealed through ^{31}P–^{1}H coupling in oligonucleotides. Pardi et al. [64] first reported this method on a single stranded nucleotide (d(CpTpApG)). Resonances in double-stranded nucleotides can also be assigned [65–67] despite the line broadening due to duplex formation.

Let me now move to the use of N.O.E. As a result of both the covalent link between monomers and the related steric hindrance, any ^{1}H–^{1}H distance between nearest neighbor units has a well defined upper limit. Wüthrich and coworkers [35, 68, 69] investigated the conformational dependence of the interresidue ^{1}H–^{1}H distances in proteins and demonstrated that one or several among these distances can be expected to be detectable in a 2D N.O.E. correlation map, whatever the protein conformation is. Similar conclusions have been drawn for oligonucleotides [35, 70] and oligosaccharides, but the assumption of a predetermined conformation (i.e. B–DNA) is here a prerequisite. Fortunately, the global assignment strategy is relatively independent of the conformation, although the tools (i.e. the observed connectivities) vary from one conformation to another. It should be mentioned that the initial sequential assignments in proteins were obtained with 1D NMR techniques [198]: conceptually, this process does not depend on 2D NMR, even if these tools make the life easier. Until now, two main traps make it difficult to transfer the tedious assignment task to an artificial intelligence. First, a cross peak may partially or completely overlap another because of the aforementioned chemical shift degeneracy (note that this pitfall is expected to be alleviated at higher fields). On the other hand, non-sequential N.O.E.'s, which originate from the particular folding of the protein (not yet known!) are also present in the spectrum and can deceive the novice... and a computer program.

Two apparently opposite, but in fact complementary, strategies have been proposed for assigning protein resonances which could be easily transferred to any macromolecule with a limited type of stereoregular structure: sequential resonance assignment [68] and main-chain directed strategy [71, 72]. In the former case, the side-chains of the various residues have to be assigned first and then the neighboring residues are identified; in the latter case, N.O.E. patterns characteristic of stereoregular structures are detected without regard to the side-chain type. Unfortunately, life is never simple and assignment is not a linear process: some information is left out temporarily and only used later and therefore the two 'theoretical' strategies are often mixed in practice.

In means of these techniques, a large number of proteins and oligonucleotides have been successfully assigned (see for example [13, 17, 73–76]), and the strategy is no longer reported in great detail despite the large amount of time spent for this tedious task.

6 General Schemes for 2D NMR of Biomolecules

A general scheme for 2D NMR includes four successive time periods: preparation, evolution, mixing and detection. In this section, various aspects that are critical for success are reviewed, and the implications of high fields are discussed.

The *preparation period* consists of a relaxation delay (T) followed by one or several r.f. pulses. In theory, thermal equilibrium should be attained during this delay; in practice, the very restrictive rule ($T > 5.T_1$) is never respected and values close to T_1 offer a better compromise from the view point of sensitivity. In many 2D experiments, where heteronuclear broadband decoupling is employed, a larger relaxation delay may be necessary because the acquisition time of the previous scan cannot be used as a relaxation period. In the molecular weight range of macromolecules that presently can be studied by NMR (MW < 15,000), typical T_1 values for ^1H are approximatively 1 s and thus the experiment can be safely repeated every 2 s or less. Even if higher magnetic fields allow the study of larger macromolecules in the future, the requirement for narrow lines implies residual internal flexibility, which will in any case contribute to reduce the T_1 value.

In the course of the *evolution period*, the individual spins evolve at their characteristic frequencies. In order to sample these, a series of experiments must be carried out with systematic incrementation of t_1. According to the sampling theorem [47], the sampling rate must be at least equal to the highest frequency contained in the signal. This means that, at higher B_0, a similar digital resolution along ω_1 (Hz per point) will require a larger number of experiments, and, consequently, a longer experimental time.

In all 2D NMR experiments, the design of the *mixing period* is the key to enhancing and selecting the information content of the spectra. Unfortunately, the selectivity goal is often achieved at the expense of experimental sensitivity and both the amount of material and machine time available are severe constraints on choosing an experiment. The mixing period is missing in *2D resolution experiments*, which aim only to separate two different spin interactions [77–79], e.g. chemical shift and J-coupling. Because of their limited disk storage requirements, homo- and heteronuclear J-resolved spectroscopy were very popular in the early days of 2D spectroscopy, but have ceased to be of great interest as high resolution correlation experiments [80] have become feasible. Depending upon the spin interaction selected by the design of the mixing period, 2D correlation experiments can be classified into two groups. In a *2D J-correlated experiment* [81, 82], a coherence is transferred between spins which share a J-coupling. In a *2D nuclear Overhauser correlation* [83, 84], the interaction involved is the dipole-dipole interaction, and connectivities are obtained between pairs, which are close to one another (< 5 Å apart).

In the course of the *detection period*, transverse magnetization is measured as a function of t_2. While the evolution and detection periods seem symmetrical at first glance, they differ in several experimental features. A large variety of spin manipulations and incrementation schemes is, in fact, allowed during the evolution period, since no NMR signal is being sampled. On the other hand, only broadband decoupling to simplify the spectrum can be performed during t_2. As a result of detection occurring *only* t_2 *and not during* t_1, the overall sensitivity is related to both abundance and γ of the spin actually detected during t_2 but only to the abundance of the monitored spin during t_1 (its γ is irrelevant). A more profound difference between the evolution and the detection periods deals with their influence on the overall duration of a 2D experiment: a two-fold improvement in digital resolution costs twice as much machine time for the ω_1 direction and is almost free of charge for the ω_2 direction. Therefore, the experimental time required for achieving the same Hz per point resolution in *both directions* is roughly proportional to the ω_1 spectral width and thus to the field.

The existence of up to 4 periods and several pulses in a 2D experiment implies the use of a selection process in order to manage the magnetization or coherence transfers along the sequence. While only z-magnetization exists prior to the first pulse, several types of coherence coexist hereafter. Thus it is usually necessary to select a given type of coherence during a period by a phase cycling process, i.e. the coaddition of several scans with different pulse phases [85]. The advance of 2D NMR has been accompanied by a proliferation of techniques for elimination of unwanted signals and artifacts, but each improvement costs at least a two fold increase in the minimum number of scans and thus in machine time. Drastic cuts in phase cycling schemes are often possible if one does not strive too much for perfection.

7 Proton 2D Spectroscopy of Biomolecules

7.1 *Chemical Shift Correlation by Scalar Spin–Spin Coupling*

One of the crucial steps in resonance assignment is the delineation of ^{1}H spin systems. Several improved pulse sequence were devised from the original two-pulse experiment (called COSY or *Jeener*-experiment) in order to reach this goal. The information content of a COSY spectrum [81] is similar to that of a set of spin decoupling experiments, but only for well-resolved spectra. 2D pulse schemes without counterpart in 1D spectroscopy have been proposed in order to cope with specific difficulties encountered in biopolymers. In fact, almost all monomers (amino acids, riboses in nucleotides or sugars) exhixit a linear spin system structure, which can be noted as A -K -M -X. Since each spin is coupled only to its two nearest neighbors, the J-link can be irreparably broken in a COSY-type spectrum for two reasons:

— if a J-coupling is too small (with respect to the line-width) to give rise to a sizeable cross-peak, no evidence linking A to the spin system can be obtained.
— a second spin system A′ -K′ -M′ -X′, corresponding to another monomeric unit) may overlap partially with A -K -M -X (for instance $\delta(K)=\delta(K')$). This situation is often encountered in proteins, where the $C^{\alpha}H$ have a much narrower spread than either the NH or the side-chain protons.

Two solutions are available to these difficulties. In the first case, an N.O.E. connectivity may replace the missing J-cross-peak. Several second-generation experiments (see Fig. 6) (relayed COSY [86], total correlation spectroscopy [87] or multiple quantum spectroscopy [88]) provide clever alternative routes to assignment. The information thus obtained is essentially complementary to that from COSY-experiments, and cannot be obtained by any 1D experiment.

Before a discussion of the various experiments and their specific purpose, it should be remembered that all classical or semiclassical vector models *fail* to describe correctly the coherence transfer achieved by a 90° pulse such as in COSY. Because the coherence in a coupled n-spin system evolves in a 2^{2n} dimensional space, our familiar 3D space is obviously inappropriate. A product operator formalism has been developed by several groups in order to analyze arbitrarily complex pulse experiments, but a description of this formalism is beyond the scope of this review, so the interested reader should refer to the original work [89–91]. Because this formalism relies on the assumption of weakly coupled spins, only computer simulation is able to manage strong coupling effects [92, 93] (see below).

If spin A and X are scalar coupled, a pair of cross peaks at frequencies (ω_A, ω_X) and (ω_X, ω_A) appear in the 2D COSY spectrum. Unfortunately, these peaks have antiphase character in their multiplet components and furthermore the diagonal peaks are dispersive in both dimensions. It can be computed that in a *weakly*

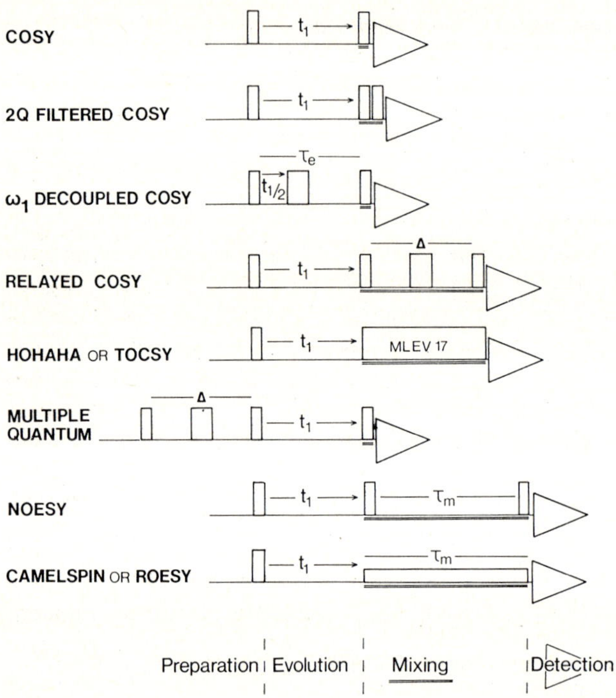

Fig. 6. Various pulse schemes used for proton 2D NMR of biomolecules

coupled two-spin system, the cross and diagonal peaks would have the same intensity only for infinitely narrow lines. As shown in Fig. 7 due to their antiphase nature, positive and negative signals partially overlap and thus cancel, leading to a decrease in intensity of the cross-peak. No cross-peak will be left if the J-coupling is smaller than the line width. Two important conclusions can be drawn from these inherent features: no change will be brought by higher field spectrometers because the J-couplings and (to some extent) the line widths do not change with field strength. On the other hand, COSY experiments on larger molecules will be possible, but only if their internal flexibility can still contribute to resonance narrowing.

Two derivative experiments (the *double quantum filtered COSY* (2QF or DQF) [94] and the ω_1-*decoupled COSY* [95, 96]) carry the same type of information as COSY, but with improved spectral characteristics. In the first case, the dispersive character of the diagonal (with respect to the cross-peaks) is removed [97], permitting display of the 2D map in the phase sensitive mode [80]. Moreover singlets (for example from residual solvent), which only would contribute to the diagonal, are removed. The double quantum filter is now a standard experiment in the arsenal of the spectroscopist, and higher order filters may also facilitate peak recognition in crowded regions of protein spectra [97]. Cross peaks associated with a particular coupling network, that is, a particular type of residue

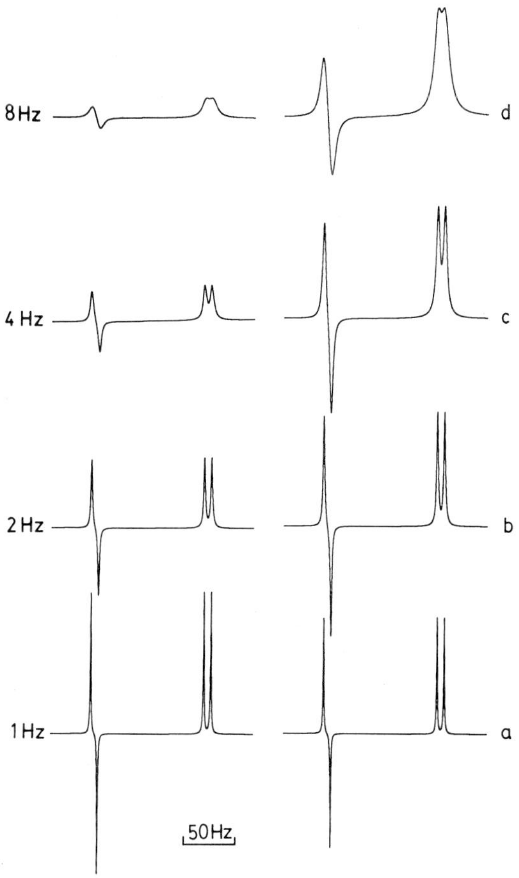

Fig. 7 a–d. Cancellation of anti-phase cross-peaks due to line broadening. In-phase and anti-phase doublets (J = 7.5 Hz) were simulated for various line-widths. The left-hand side spectra have the same scaling, and the right-hand ones were scaled on the in-phase doublet. The intensity of the anti-phase cross-peaks decreases much faster than that of the in-phase

such as *Gly*, may be eliminated by this technique, although it suffers from inherently low sensitivity. On the other hand, the ω_1 decoupled—or constant time—COSY [95, 96] leads to a gain in resolution in the ω_1 direction and thus may remove ambiguities caused by overlapping multiplets. The multiplet structure in the ω_1 direction collapses (as in ^{13}C NMR by means of broadband decoupling) for lack of J-modulation during the constant evolution period. This experiment, tuned for a precise value of the scalar coupling ($J = (2\tau_e)^{-1}$) is more "susceptible to misadjustment of experimental parameters (τ_e ..)" [96] than, for example, the DQF COSY.

Early 2D spectra were displayed in the absolute mode because of the dispersive character of the COSY diagonal and also an experimental limitation (minimal delay between pulses with different phases). Except for particular pulse sequences which inherently prevent recording absorption spectra (such as phase modulated experiments), absolute value mode spectra completely waste the spectral resolution achievable on high field spectrometers as shown in Fig. 8.

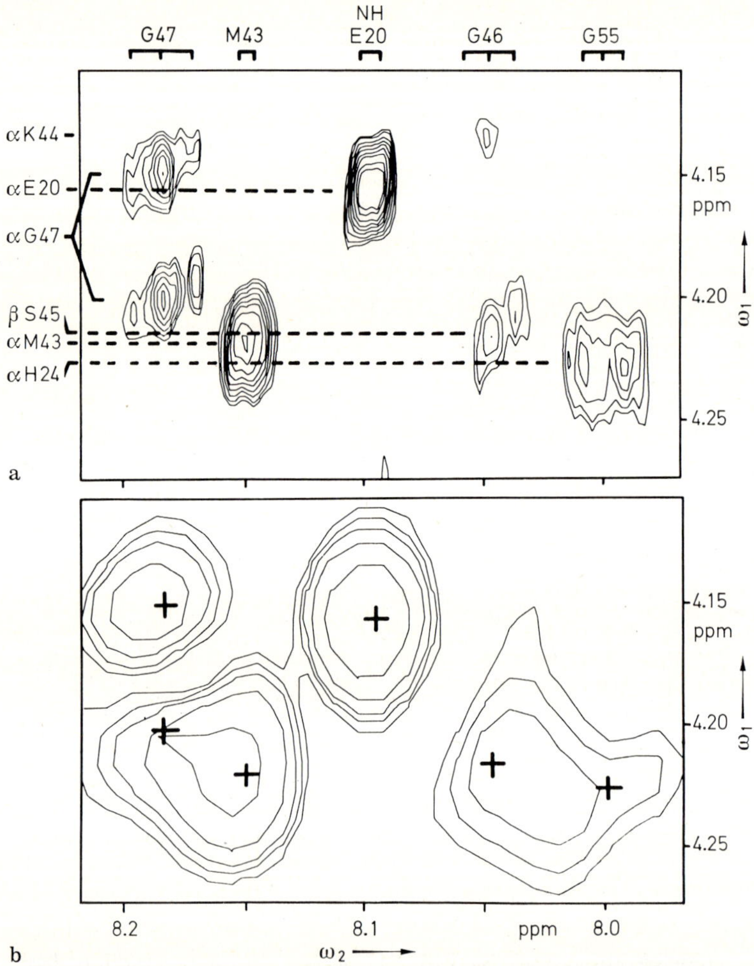

a b

Fig. 8 a, b. Absolute value displays versus phase-sensitive (or absorption) displays. a) High resolution absorption mode display of a region of the NOESY spectrum of proteinase inhibitor II A from bull seminal plasma (resolution 2.4 Hz in ω_1 and 0.6 Hz in ω_2) b) Low-resolution absolute value display of the same region (resolution 9.6 Hz and 4.8 Hz respectively). *Crosses* mark the centre of the peaks. The apodization functions were optimized independently in both cases. Reprinted with permission from Williamson, Marion and Wüthrich [114]. Copyright Academic Press Ltd (1984)

From the COSY (or preferably from the DQF COSY) of a biopolymer, two major pieces of information can be obtained: identification of the spin system (i.e. the chemical nature of the corresponding fragment) and quantification of the involved J-coupling (a conformational probe). As far as proteins are concerned, the first piece of information amounts to recognizing a particular geometric pattern (either in H_2O or D_2O spectra) which is characteristic for each of the different spin systems encountered in the 20 common aminoacids (see ref [35]

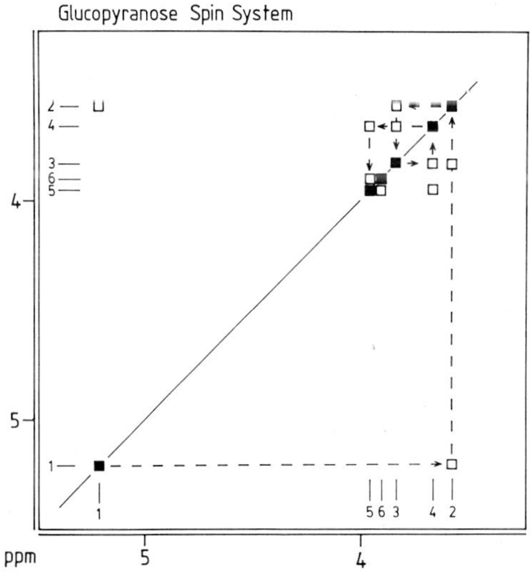

Fig. 9. The glycopyranose spin system. The chemical shift are taken from the α-anomer of Glc in α-Neu5Ac-(2–3) βGal-(1–4) Glc (see Fig. 3) reported in [162]. Note that the chemical shifts of most protons are between 3.5 and 4 ppm, which implies the use of an high field spectrometer

p 136 for a survey). Some residues, such as *Gly* or *Thr*, have a unique pattern but others such as *Phe*, *Asp* or *Ser* have a similar topology and will never be distinguishable even at higher field. A similar approach can be applied to oligosaccharides and derivatives [41, 98, 99] in spite of the severe resolution problem peculiar to this type of molecule ("the hidden resonance problem") (refer to Fig. 9). Provided that the complete J-connectivity can be extracted, which will be made easier at higher fields, an aldopentose can be in principle distinguished from an aldohexose: a 6-spin system is topologically different from a 7-spin one.

The resolved fine structure in *phase-sensitive COSY* [80] can be of great help for obtaining accurate J-coupling measurements. The 2D-multiplet pattern is often complicated by the effect of finite line-width and limited digital resolution. While the first cause is nearly independent of the field strength, the large spectral width required at higher fields may become a major obstacle in J-measurements. Moreover, due to the anti-phase nature of the J-cross peaks, overlap causes partial cancellation and consequently the apparent line separation is larger than the actual J-coupling [100].

Without discussing in detail the conformational use of the J's, let me give two examples. Helical structures in proteins can be identified readily by small value of $^3J_{NH-C^\alpha H}$ (4 Hz) while extended structures (either parallel or antiparallel) exhibit larger couplings (9 Hz). Similarly, the configuration and the conformation of aldopyranose residues can be established by means of qualitative estimation of several vicinal coupling constants (41): however, the absolute stereochemistry (L- or D-form) cannot usually be determined by NMR.

Let us now focus our attention on second generation 2D experiments, starting with relayed COSY (Fig. 6). In a linear 3-spin system AMX with $J_{AX} = 0$, an A-coherence is transferred to M (by the second 90° pulse) and next to X (by the third 90° pulse). Thus one has at one's disposal a very powerful assignment technique *without counterpart in 1D spectroscopy*. Although the M resonance may be overlapped and thus hidden, this technique can reveal its location. Even at high field, this is a rather common situation. Note that these spectra should be interpreted with caution, because the fixed delay, actually tailored to J's close to $(2\Delta)^{-1}$ [101], may lead to accidental vanishing of cross peaks for other J-values. Therefore, in work with macromolecules showing a large range of J-coupling [102, 103], it is advantageous to aim for a compromise in the delay selection, since repetition of the experiment with different delays is likely to be much too time-consuming.

The basic idea of relayed transfer has been extended by including a series of n consecutive sequences (90° $\Delta/2$ 180° $\Delta/2$) in order to transfer coherence across n couplings in a linear spin system in order to resolve ambiguities in assignments. This approach has been applied to BPTI, a 58-residue protein [102], and a pentasaccharide [104]. Let finally point out two main drawbacks of the relayed COSY: the choice of Δ values requires preliminary knowledge—or at least an educated guess—of typical J's, and the diagonal and cross-peaks are 90° out-of-phase (this can be resolved by multiple-quantum filtering [105, 106] at the expense of the sensitivity).

An elegant alternative to the multistep COSY is provided by *total correlation spectroscopy* (TOCSY or HOHAHA [87] depicted in Fig. 6). During the mixing period, a *net* coherence transfer can actually be achieved by several multipulse schemes, the purpose of which are the complete elimination of chemical shifts. The coherences migrate in an oscillatory manner throughout the entire spin system. Cross peaks are expected between all nuclei belonging to the same spin system, even if not directly coupled. The simplest mixing scheme employs a string of 180° pulses, the repetition rate of which should be faster than the largest chemical shift difference (typically 3000 to 5000 Hz for a protein at currently available field strengths)[1]. Only direct connectivities are observed for short mixing times ($< 0.1/J$). The complete spin system can be detected with a longer mixing time (20–100 ms) provided that no small coupling break off the link. This method has been successfully applied to oligosaccharides [108] (a complete subspectrum is obtained for each monosaccharide), proteins [75] or oligonucleotides. In each case, one reaps the benefit of the larger dispersion in chemical shift of one proton (the amide proton in proteins or the anomeric proton in saccharides) in order to map the entire spin systems.

The number of reported HOHAHA spectra was until recently limited by hardware limitation and by the fear of some spectroscopists of moving from

[1] Better results can be obtained by means of pulse sequences which compensate for the various imperfections (phase and spatial inhomogeneities). Sequences proposed for heteronuclear broadband decoupling (MLEV ...) [107] are candidates for achieving isotropic coherence transfer, but removing a ^{1}H chemical shift (several kHz) is much harder than removing a ^{1}J coupling of 100–150 kHz. Improved pulse sequences will be required in order to match the increased chemical shift range

the easy COSY (in which only a few pulses are used) to the Hartman–Hahn experiment (where a great number of pulses are applied). Nonetheless, it is likely that this situation will evolve, so that the latter experiment will supplant any COSY-related scheme in the near future.

Multiple quantum (MQ) spectroscopy is able to trace out extended coupling networks [88, 109]. The most exciting prospects for use of this technique arise from the fact that, in addition to information on remote connectivities and spin equivalence [88], evidence of coupled spins is given by a geometrically different pattern: in a two-spin system, the double quantum spectroscopy gives rise to peaks at $(\omega_A+\omega_X, \omega_A)$ and $(\omega_A+\omega_X, \omega_X)$. Therefore, partial peak overlaps occur in a different way and can be recognized more easily. Furthermore, due to the different geometry, one may also expect to observe cross peaks which would have disappeared in a standard COSY as a result of water saturation (see below). (see Fig. 10) No "bleaching" of cross peaks [110] can occur in the ω_1 direction,

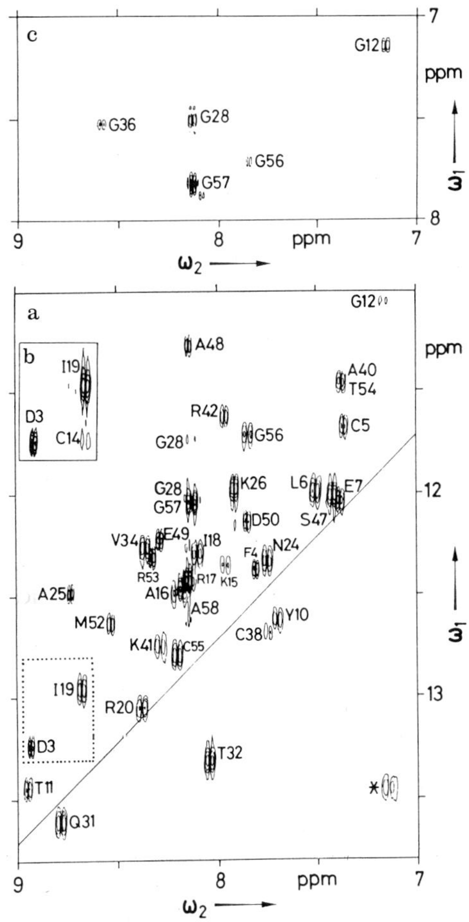

Fig. 10a–c. Phase-sensitive double-quantum 2D NMR spectrum of a 20 mM of BPTI in 90% H_2O/10% D_2O at 35.5 °C and pH = 2.1. **a)** Direct connectivities: note that R20 cross-peak was completely bleached out in the DQF COSY spectrum shown in ref. [110], because the $C^\alpha H$ was almost under the saturated water line. **b)** Plot of the area inside the dotted lines at a lower level. **c)** Remote connectively peaks for glycine because of the two $C^\alpha H$. Reprinted with permission from Otting and Wüthrich [110]. Copyright Academic Press Inc. (1986)

since no double quantum coherence of H_2O can be created. The advantage arising from the *absence of diagonal peaks* becomes substantial when the cross peaks get closer to the diagonal (relatively strongly coupled spins). As mentioned above for other experiments, the interval Δ also must be tuned to a preselected value of J-coupling, but improvements over this conventional experiment can be achieved with symmetrical excitation–detection schemes [111]. For assignment purposes, it is undubitable that the information content of MQ spectrum is complementary to that of standard COSY spectra if the latter is recorded at higher field. The range of macromolecules amenable to analysis can be extended, at a constant field strength, by MQ spectra as recently demonstrated on relatively large proteins (MW = 15000 – 18000) [112]. In addition to the assignment of numerous residues, MQ spectroscopy has made it possible to detect long-range connectivities between aliphatic and aromatic resonances in *His* [113].

7.2 Chemical Shift Correlation by Nuclear Overhauser Effect

So far we have dealt only with J-correlation spectroscopy. In all reported methods, we have seen correlations via non-vanishing scalar spin-spin coupling acting as the vehicle for transferring *coherences* among coupled spins. There is yet another important realm of 2D correlation spectroscopy wherein internuclear transfer of *magnetization* takes place via an incoherent process, the nuclear Overhauser effect [119, 120]. Although the incoherent character of this interaction strongly limits the spectroscopist's imagination in devising new pulse sequences, this experiment is of central importance in the study of macromolecular conformations.

The nuclear Overhauser effect (related to cross-relaxation) depends in part on modulation of the dipole-dipole interaction between different nuclei by the Brownian motion of the macromolecule. Therefore its intensity depends on the character of the motional process *and* on the separation of the interacting protons:

$$\text{N.O.E} = 1/r^6 * f(\tau_c)$$

As far as the time scale of the motion is concerned, we have to distinguish the *fast-motion limit* ($\tau_c \ll \omega_0^{-1}$) which corresponds to small molecules and the *slow-motion limit* ($\tau_c \gg \omega_0^{-1}$) which corresponds to macromolecules studied at high field. It is apparent that the slow-motion limit will be the standard situation as higher field spectrometers become available.

Early N.O.E. measurements on small molecules were carried out by *steady-state methods*, in which one spin was selectively saturated while the intensity change occurring on its relaxation partners was observed [120]. However, in multi-spin systems where the magnetization can transfer *via* relays, the N.O.E.'s have to be used with caution for distance quantification. A reliable quantitative interpretation requires a rigourous mathematical treatments involving all interacting spins. However it should be mentioned that, in the case of fast motion, the information is blurred but not irretrievably lost by multi-spin effects.

For studing macromolecules at high field (slow motion limit), steady-state 1D methods are impractical because of the partial resonance overlap *and* extensive spin diffusion. In fact, spin diffusion is just another name for multi-spin effects when they become a more severe problem in macromolecules [121]. Because of the imbalance of the self relaxation rate relative to the cross relaxation rate, magnetization can actually transfer much faster from one spin to another than it can decay by leakage to the lattice. Consequently, specific information on distances in biopolymers can only be obtained by *transient methods*, where the magnetization transfer among spins is studied after the initial perturbation of one of them.

The 2D adaptation of the transient N.O.E. is the NOESY experiment (Fig. 6) [83, 122] which turns out to be more suitable for the slow motion limit than for the fast one [84]. For sufficiently short mixing period (τ_m), sizable positive cross-peaks are detected. In the conversion of cross-peak intensities to interproton distances, several pitfalls must be skirted to avoid misinterpretation of NOESY spectra.

In a NOESY experiment, also referred as to *2D exchange spectroscopy*, effects arising from incoherent magnetization transfer either by N.O.E. or by chemical exchange are registred. Unfortunately, in the slow motion regime both N.O.E. and chemical exchange cross peaks are positive with respect to the diagonal and hence are indistinguishable. For example in side chain amide groups of the peptidic residues *Asn* or *Gln*, the intensity of the cross peak between the two nonequivalent protons [123] cannot be ascribed solely to N.O.E. but also to the *cis-trans* isomerization about the amide bond.

From comparison of the NOESY pulse scheme with that employed for relayed COSY (Fig. 6), it is apparent that NOESY spectra may contain additional cross peaks (called *J-peaks*) originating from coherent transfer of magnetization during τ_m. Due to the second 90° pulse, a large variety of magnetization (z-magnetization) and coherences (zero-quantum (ZQ), one and multiple quantum) are evolving during τ_m. All but the ZQ coherences can be suppressed by proper phase cycling [124]. Removal of ZQ coherence has to exploit their oscillatory evolution during τ_m.[2] In the spin diffusion limit ($\omega\tau_c \gg 1$), the relaxation rate of ZQ coherence and the cross-relaxation responsible for the cross peak build-up are identical [124]:

$$1/T_2^0 = \sigma = (1/10)(\gamma^2 h^2 r^{-3})^2 \tau_c$$

for a two spin system. Choosing a mixing time such that the ZQ coherences would have disappeared due to transverse relaxation amounts to deliberately waiting until spin diffusion dominates the NOESY map.

[2] Various strategies have been proposed [125] but none is completely satisfactory: incrementation of the mixing time parallel to t_1, random variation or addition of a 180° pulse at random position. However, an optimal suppression by such techniques may be incompatible with available machine time, because of the number of scans required

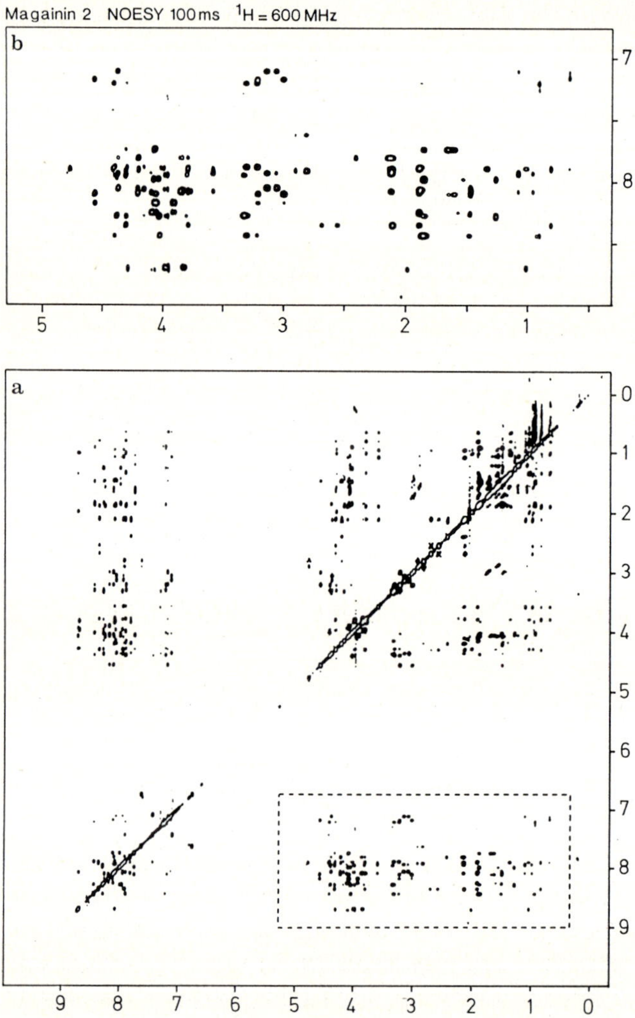

Fig. 11 a, b. 600 MHz NOESY spectrum of magainin-2, an antimicrobial 23-residue peptide isolated from skin of the african frog *Xenopus Laevis* [126]. 15 mg were dissolved in a mixture of H_2O-trifluoroethanol (70:30 v/v). All NH(i)–NH(i + 1) connectivities and most of the $C^{\alpha}H(i)$–NH(i + 3) (inset **B**) are observed, evidence of the helical conformation (see Fig. 1) adopted by this peptide under these conditions

Once the resonances of a protein have been assigned using N.O.E. in a qualitative manner as a link between residues, the sequential N.O.E. (i.e. involving nearest neighbor residue) combined with other medium- and long-range N.O.E.s allow characterization of the elements of stereoregular secondary structure. Figure 11 shows the 600 MHz NOESY spectrum of magainin-2, an antimicrobial peptide of 23 residues. Its conformation, which is an α-helice

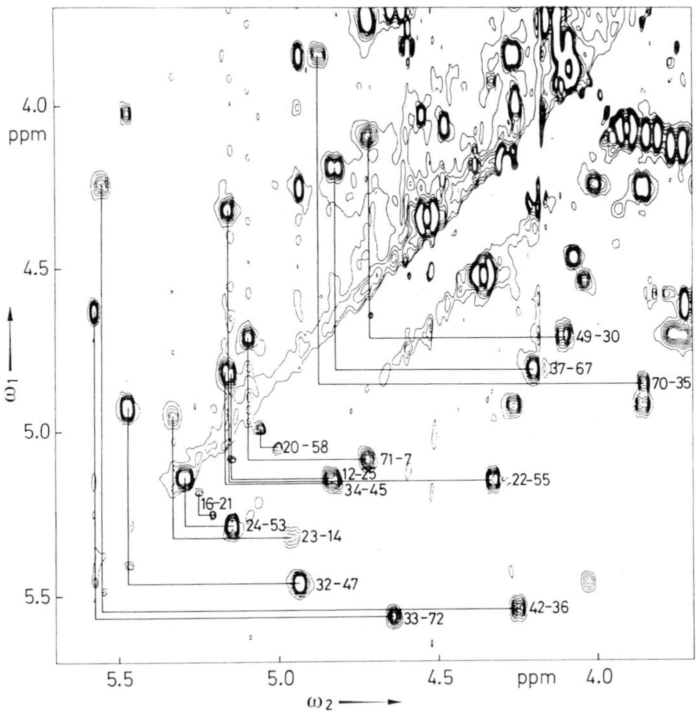

Fig. 12. Region of the ^{1}H NOESY of Tendamistat containing the $C^{\alpha}H$–$C^{\alpha}H$ cross-peaks (concentration 5 mM, mixing time 200 ms, diagonal suppression). The numerous long-range N.O.E. evidence two anti-parallel β-sheets, each made of 3 major strands (see Fig. 13). High field (here 500 MHz) is a requirement, since these connectivities are close to the diagonal. Reprinted with permission from Kline and Wüthrich [128]. Copyright Academic Press Ltd (1985)

running to both ends of the peptide [126] is definitively and easily established by two continuous stretches of N.O.E.: one stretch between adjacent NH's (d = 2.8 Å) and a second one between the $C^{\alpha}H$ of residue (i) and NH of residue (1 + 3) (a turn of an α-helice entails a bit more than 3 residues). In Fig. 12, a detail of the 2D NOESY spectrum of Tendamisat, a 74-residue α-amylase inhibitor, is shown [127] and the arrangement of the polypeptide chain in β-sheets is manifested by short $C^{\alpha}H$–$C^{\alpha}H$ distances (< 4.0 Å) between residues far apart in the sequence. These connectivities enabled locating one strand with respect to another (Fig. 13) and evidence was found that 3 β-bulges prevail near residue 43 to 44, 56 to 57 and 68 to 69. Comparison of this solution structure with the X-rays conformation obtained independently shows that the interior of Tendamistat is nearly identical in both states, whereas a small number of differences have been detected at the surface of the protein [128].

The interaction of macromolecules can be monitored by NOESY spectra. The appearance of internuclear N.O.E.'s provides not only proof of complex formation but also very specific structural data. An illustrative example is given

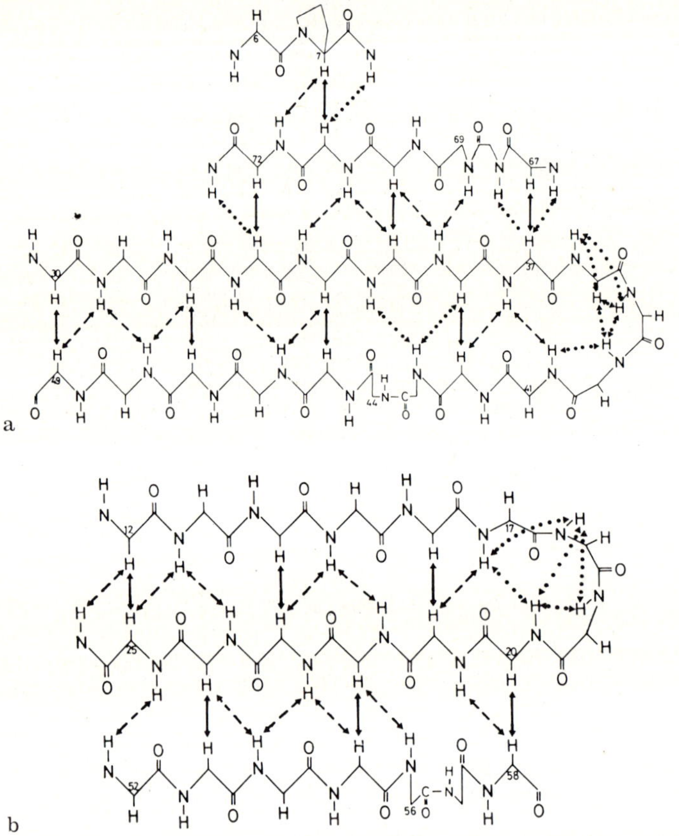

Fig. 13 a, b. Antiparallel structure in Tendamistat as determined of the following connectivities: *Solid arrows* $C^{\alpha}H(i)$–$C^{\alpha}H(j)$ (refer to the experimental spectrum of Fig. 12), *broken* and *dotted* arrows $C^{\alpha}H(i)$–NH(j) and NH(i)–NH(j). The connectivities noted by dotted arrows are only observed in H_2O and thus correspond to relatively fast exchangeable protons. Reprinted with permission from Kline and Wüthrich [128]. Copyright Academic Press Ltd (1985)

by the study of the interaction of the *lac* repressor headpiece with either a 14 or a 22 basepair *lac* operator fragment (see Fig. 14) [129]. Analysis of the complete spectrum is complicated by strong overlap of intraprotein and intra-DNA cross peaks, but on the basis of assigned N.O.E.'s between both partners, a model can be built (Fig. 14b). The main conclusion of this outstanding investigation is the orientation reversal of the protein with respect to DNA, when compared to previously available models [130]. Although one might argue that this difference originates from the use of a fragment, recent genetic studies [199] confirm this result on the whole repressor.

I will now focus on the problems raised by the use of higher field spectrometers. According to the above equation, the build-up rate of a NOESY cross-peak is independent of the Larmor frequency in the spin diffusion limit. Therefore, moving to be suppressed to a higher field on the same

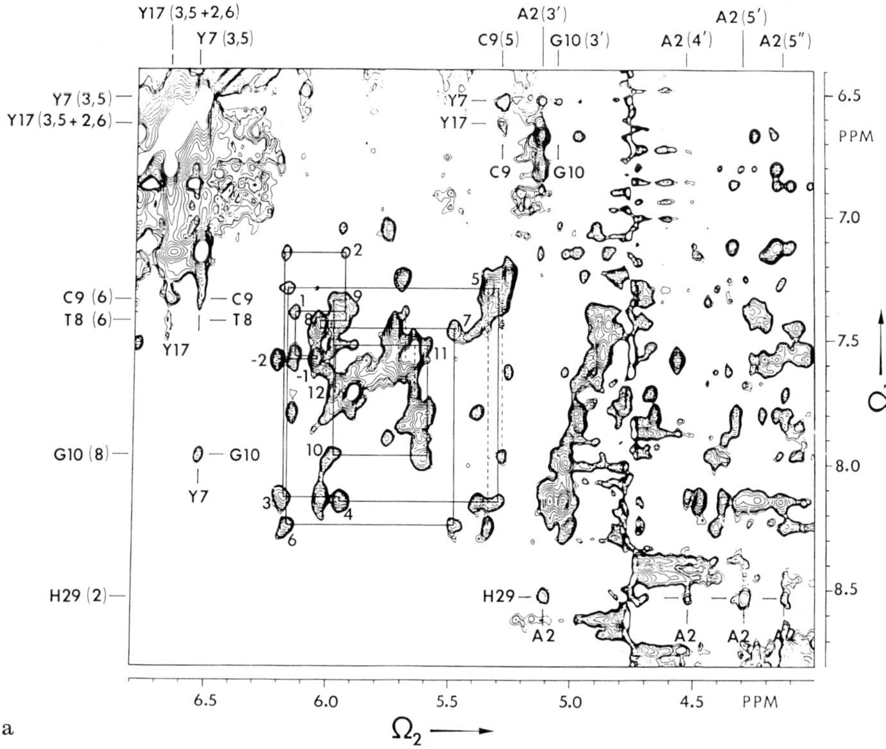

a

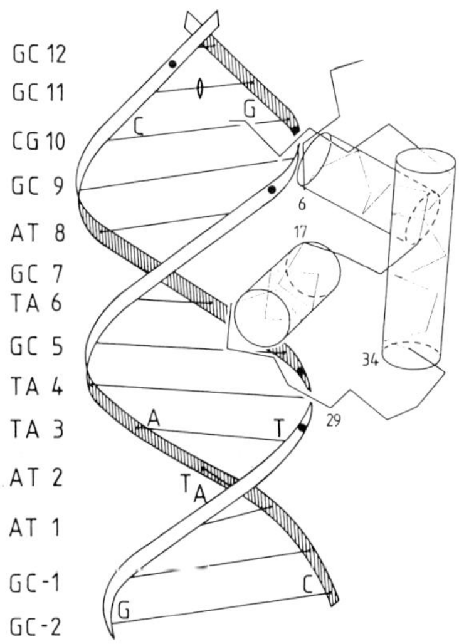

b

Fig. 14 a. 500 MHz NOESY spectrum of the complex of *lac* repressor headpiece with a 14 base-pair operator (5 mM in both protein and DNA, pH 7, 27 °C). The sequential N.O.E. between base protons (purine H8 and pyrimidine H6) and ribose H1′ are reported for one strand of the DNA. N.O.E. cross-peaks indicated between protein and DNA allows one to build the model shown in **b**. (the DNA is a standard B–DNA and the headpiece structure is that of the free protein). Reprinted with permission from Boelens et al. [129]. Copyright Academic Press Ltd (1987)

macromolecule will make the study easier without adding troublesome features to its NOESY spectra. On the other hand, higher fields will permit conformational studies of larger molecules, which are likely to have longer overall correlation times. In the absence of internal flexibility, cross-peaks will build up faster and multi spin effects will become more efficient even at short mixing times (< 50 ms). A more severe limitation is the concommittant broadening of the line width, which also increases proportionally to τ_c.

If there is no possible change of the motional parameter (τ_c), the only way to ameliorate the line-width problem is to decrease proton relaxation by removing a fraction of protons (the averaged H–H distance is extended). Much narrower signals can be restored by diluting the local proton density by means of a random fractional (roughly 75%) deuteration of the molecule, as recently reported by LeMaster and Richards [131] on a 108-residue protein. The line width, which originates from the contribution of all relaxation partners in the immediate vicinity of the observed one, is reduced; but the initial build-up rate NOESY cross peaks (where only a single pair of proton is concerned) remains unaffected. However, as a global feature resulting from all pairs, the spin diffusion contributes less to the NOESY spectra because of spin dilution.

To compensate for the lack of a conventional N.O.E. near the condition $\omega_0\tau_c = 1$, an experiment analogous to NOESY has been proposed for the measurement of rotating frame Overhauser effects under a spin-locked condition [132, 133]. In this experiment, initially named CAMELSPIN and then ROESY, magnetization transfer occurs along a spin lock field B_1 which is several orders of magnitude smaller than the magnetic field B_0. Whatever the rate of the molecular tumbling is, *a positive N.O.E.* is observed. The ROESY technique is presently a very attractive technique for intermediate sized molecules such as 8–10 residue peptides (an example is given in Fig. 15) [134] or oligosaccharides [132], which unfortunately meet the condition $\omega_0\tau_c = 1$ under typical conditions of solvent and temperature.

One cannot gloss over multi-spin effects which (as for conventional N.O.E's) are likely to corrupt the data interpretation. However, as a result of the positive sign of the rotating-frame N.O.E., the selectivity of information is not damaged to the same extent as when spin-diffusion occurs in the conventional N.O.E. experiment. At first glance, ROESY looks promising for conformational study of large macromolecules, bypassing the painful NOESY spin-diffusion limitation. This attractive promise ignores the fact that the rotating frame is governed by a T_2 relaxation process [132]. For slow molecular motion, the cross-relaxation rate is:

$$\sigma_2 = (1/5)(\gamma^2 h^2 r^{-3})^2 \tau_c$$

whereas the self-relaxation rate is:

$$\rho_2 = (1/4)(\gamma^2 h^2 r^{-3})^2 \tau_c$$

Before reaching detectable intensities, the small N.O.E. peaks will be washed away by self-relaxation. The distance range accessible by ROESY is shorter than

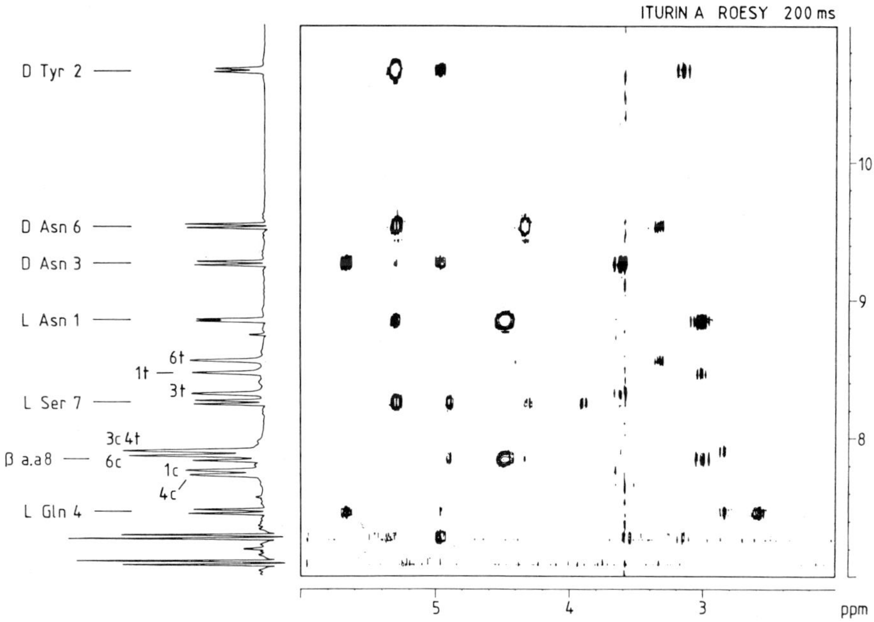

Fig. 15. 300 MHz ROESY spectrum of iturin A, an antifungal lipopeptide from *Bacillus Subtilis* [123] (Solvent pyridine-d_5. 27 °C). Because of the rotational correlation time of this molecule (8 peptidic residues), no N.O.E. are detectable in a NOESY spectrum at room temperature

achievable by NOESY. In spite of repeated warnings concerning spin-diffusion in NOESY, one thoughtlessly reaps the benefit from the slow T_1 self relaxation in the spin diffusion limit in order to reach distances up to 4 or 5 Å. From this view point, at least one advantage of ROESY is that overinterpretation of the data is not possible.

Conceptually, the ROESY experiment shows an obvious family likeness to the TOCSY or HOHAHA experiment dicussed previously. The major difference lies in the relatively stronger r.f. field required for achieving J-transfer in HOHAHA. It is therefore not surprising that, as discussed for NOESY, efforts have been devoted to suppressing J-transfers in ROESY maps. Efficient techniques have been proposed for this purpose including a fast sequence of small flip angle pulses [135] or relatively weak off-resonance continuous wave irradiation [133, 136].

7.3 *Observation of Labile Protons: Suppression of Water Resonance*

By definition, a labile proton has a limited life time on a molecule and will be replaced sooner or later by a proton of the solvent molecule, usually water. Labile protons (for instance *amide protons* in proteins or *amino and imino protons* in

DNA fragments) are able to form hydrogen bonds which stabilize the conformation and are often at key locations in the structure. Their observation requires the recording of spectra in H_2O, so one faces the severe dynamic range problem of detecting weak resonance signals (typically several mM) in the presence of a strong solvent line (110 M). For convenience, let me divide the exchangeable protons into three classes according to their life time: slow exchangeable ones (several minutes or even hours), intermediate ones (about several seconds) and fast ones (10 ms up to 1 s).

2D spectra of slowly exchanging protons can be recorded on freshly prepared D_2O solutions [137]. The time dependence of the intensity of the individual COSY cross-peaks has been used in BPTI and other globular proteins for determining exchange rates of NH protons. Rapidly exchanging ones correspond essentially to the protein surface, while exchange rates slower than 10^{-1} min^{-1} [138, 139] are observed on protons shielded from solvent contact.

Observation of intermediate (such as surface NH in proteins at low pH) or fast exchangeable ones (amino or imino protons in nucleotides) can only be carried out in H_2O solution but the particular solvent suppression technique used differs from one case to the other. For intermediate rates, the water resonance can be suppressed merely by continuous selective irradiation [140] applied at all times except during the evolution (t_1) and detection (t_2) period. By optimization of the r.f. homogeneity and power, only a very few NH–C^{α}H cross-peaks will be bleached out, and this problem can be easily circumvented by the recording of several spectra at different temperatures, provided that the conformation remains unchanged. In fact, the large temperature coefficient of the chemical shift of water is an advantage here, whereas it is usually a disadvantage causing the spectra to be very sensitive to temperature fluctuations via the lock on D_2O. Irradiation during T_1 leads to Bloch-Siegert shifts and distorted spectra [141]. A further improvement in spectral quality can be obtained if the water irradiation is done using a r.f. field coherent with the carrier frequency [142].

Relatively fast exchangeable protons cannot be detected using presaturation, since they are likely to be replaced by a saturated proton from H_2O by chemical exchange during the presaturation delay. As an alternative, pulse sequences have been devised first for 1D NMR [143] and then extended to 2D, which excite only the desired spectral region, leaving the solvent magnetization oriented along the static field direction. Soft pulses have been used in the past, but excitation with specifically designed composite hard pulses is less sensitive to r.f. inhomogeneity. The degree of suppression varies with the technique selected; but in any case, one can not dispense with fine adjustment of the pulse lengths, delays and carrier frequency. One of the remarkable successes of these techniques has been the observation of imino protons in short DNA fragments or tRNA. As a consequence of their remote chemical shift (12 to 14 ppm), they can be selectively excited and assigned by means of 1D N.O.E.'s with exchangeable or nonexchangeable protons of adjacent bases [144, 145]. The exchange rates of the imino protons, which are catalyst-sensitive, provide insight into the non-cooperative mechanisms of base-pair opening, provided that the study is correctly done *in presence of a large excess of catalyst* [145]. In 1D spectra, the baseline roll due to dispersive wings of the water line rotated by large phase correction is a minor problem that can be avoided by iterative baseline correction, but this will be a major concern in 2D NMR.

In the earlier adaptation of the selective excitation technique to NOESY, only the final pulses was replaced by a selective pulse. The imino proton spectra of $tRNA_1^{Val}$ [146] and $tRNA^{Phe}$ [147] have been almost completely assigned by 2D methods. The 2D excitation profile thus exhibits a spectral amplitude response of the selective pulse along ω_2 but is uniform along ω_1. As a consequence of the

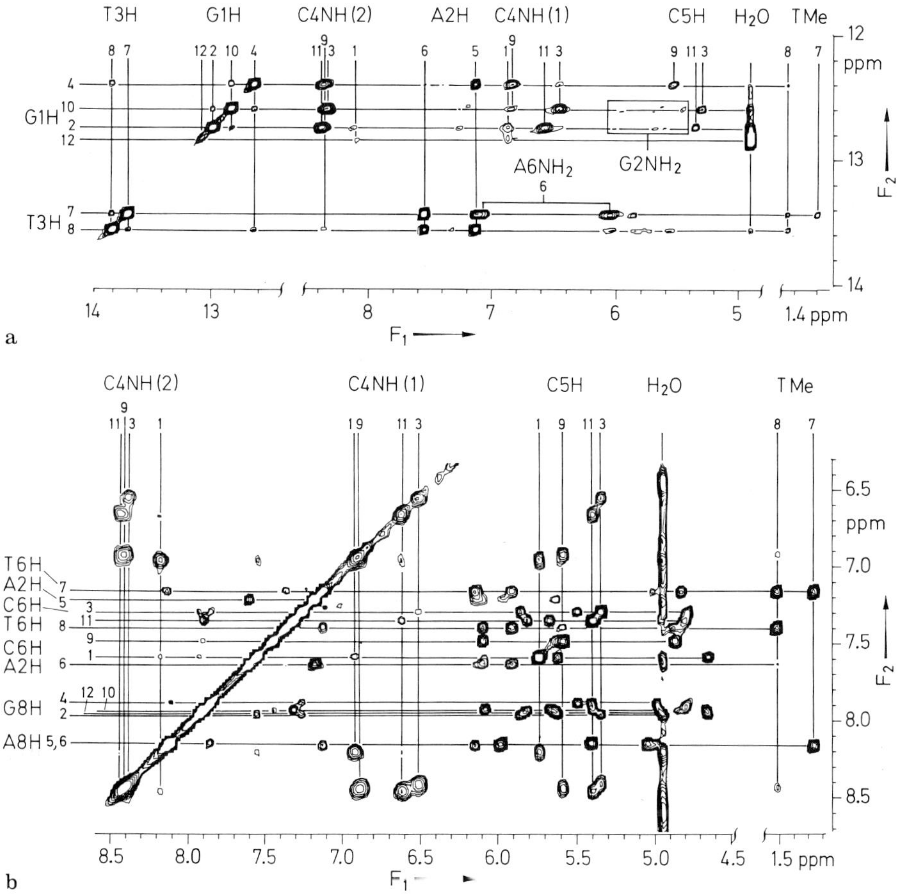

Fig. 16 a, b. Part of the 2D N.O.E. spectrum of $d(CGCGAATTCGGG)_2$ (the "*Dickerson dodecamer*") in H_2O at 5 °C with 100 ms mixing time. **a)** interaction with the imino protons. **b)** interaction with the amino and non-exchangeable protons. Only the last pulse of NOESY was replaced by a selective excitation and any x–y magnetization during the mixing period is destroyed by a 30 ms homogeneity spoiling pulse. The data were obtained on 500 A_{260} units of dodecamer in 15 h. Reprinted with permission from Sklenar at al. [150]. Copyright Fed. of Eur. Biochemical Societies, Elsevier Science Publisher B.V. (1987)

phase cycling required for ω_1 quadrature detection, water magnetization is brought into the *x-y* plane during the mixing time of some scans, to that a huge water signal must be sampled. This can be solved in two ways: a *homogeneity spoiling pulse* can be applied for several tens of ms during the mixing period [148] or all three 90° pulses of the NOESY sequence can be replaced by selective pulses [149]. For studies of oligonucleotides, either solution is appropriate [149, 150], but for proteins, the complete symmetry of the excitation profile with 3 selective pulses will make some NH–C^{α}H cross peaks disappear completely in both dimensions. It should be remarked that a further improvement in the spectrum quality can be achieved by phase cycling, as recently demonstrated on a DNA dodecamer shown in Fig. 16 [150].

In conclusion, the observation of exchangeable protons in biopolymers at mM concentration is now feasible by 2D NMR techniques. Presaturation and selective excitation techniques will both benefit from the larger spread of chemical shifts brought by the next generation of high field magnets.

7.4 Strong Coupling in ^{1}H Correlation Experiments

The two major outcomes of limited available field strength are first, the partial or complete overlap of resonances and second, the occurrence of strong coupling (that is, a J coupling becomes of the same order of magnitude as the chemical shift difference). Spectral distortion caused by strong coupling makes the data interpretation difficult and in some cases impossible. In 1D NMR spectra, strong coupling effects appear in the modified frequency and amplitude of multiplet components, not only in the spectrum of the strongly coupled pairs, but also in that of any spin which shares a J-coupling with them. The reliable interpretation of these spectra require the use of computer program devised either for fitting or for simulating them.

Similarly, 2D J-correlation spectra are also influenced by strong coupling. In these spectra (as long as no MQ coherences are excited), the frequencies along both axis correspond to that observed in 1D. Therefore, there is no doubt that the line frequencies in a 2D cross-peak will be more or less slightly altered by strong coupling, as well as their amplitude and phase. In the 1D spectrum of a AB spin system, the amplitude of the inner lines is larger than that of the outer, giving rise to the so-called *roof effect*, which proved to be of great help for assigning spectra in the 1960s. In COSY [81] or DQF COSY [151] of the same spin system, a bidimensional roof effect is observed. In exact analogy to 1D NMR, the amplitude of the individual lines are related to the strong coupling parameter θ defined as:

$$\tan(2\theta) = J_{AB}/(\delta_A - \delta_B)$$

Unfortunately, the cross-peak component, which is the furthest from the diagonal, has the smallest amplitude, and its detection (as thus the evidence of the connectivity) becomes more difficult as the coupling get stronger. However, weak or missing connectivities can be replaced for nearly-degenerate spins by proof of magnetic equivalence available from a MQ spectrum [109]. The complexity of strongly coupled 2D spectra originates from the break-down of several rules [47] which simplify other spectra: refocusing 180° pulses (used in relayed COSY [152], for instance) will no longer eliminate chemical shifts [47]. On the other hand, in an ABX system with $J_{AX} = 0$, a coherence transfer can occur between A and X, so that the appearance of a cross-peak at (ω_A, ω_X) in COSY cannot be taken as proof of the existence of a non-vanishing J_{AX}.

In conclusion, several features in 2D strongly coupled spectra offend the common sense of the spectroscopist. The safest way of interpreting them resorts to simulation programs capable of managing these specific features. Such a tool is provided by the simulation program named SPHINX [92, 153], which explicitly calculates homonuclear (and heteronuclear) 2D spectra with strong couplings. It relies, not on the product operator formalism [89–91], which here fails, but on the density matrix formalism [81]. The spin evolution along the pulse sequence is computed for the various phase steps and then coadded. Using this versatile and powerful tool, Widmer and Wüthrich [92, 153] simulated the cross-peaks encountered in proteins or DNA in which strong couplings are likely ($C^{\beta}H_2$ and $H^{2'}H^{2''}$ respectively). Digital resolution strongly influences the look of 2D spectra and it is thus convenient and reliable to compare experimental and simulated spectra under the same conditions. Figure 17 shows the simulation of the cross-peaks between the $H^{2'}H^{2''}$ and the $H^{3'}$ and $H^{1'}$ respectively. The deoxyribose ring is known to exist in two preferred conformations [40] *C3′ endo* (which prevails in A DNA and for guanosine in Z DNA) and *C2′ endo* (in B DNA and for cytosine in Z DNA) (see Fig. 2). Assuming a rapid interconversion between A and B DNA conformations, the spectra have been simulated for various populations for comparison with experimental data. This example emphasizes the fact that the information yield of a 2D J-correlation spectrum can be amplified by simulation, when the field strength becomes a limiting factor. The future of the data interpretation (whether strong coupling are present or not) relies on fitting (and not only simulating) the experimental data versus the spectra parameters and later versus the conformational one.

Let me now address the problem of the behavior of the N.O.E. in strongly-coupled spin systems. For a long time, strong coupling effects have been known to cause T_1 and T_2 to deviate from classical behavior (see [53] for a review). Thus N.O.E.s, which originate from dipolar relaxation, are likely to be affected as well. Consequently, more or less serious errors in interpretation can arise from disregarding strong coupling effects [154, 155]. In βD Man*p*1-OMe, a monosaccharide, a 2% N.O.E. between H_1 and H_4 was reported by Brisson and Carter [155]. This effect has been ascribed to the strong coupling between H_3 and H_4 ($J/\delta = 0.4$), rather than solely the result of multi-spin effects. Recently, two different groups [156, 156] have investigated the effect of strong coupling on 2D NOESY maps. From early developments in NMR, it is known that it is no longer possible to ascribe a particular line to an individual spin when strong coupling arises and thus N.O.E.'s lose their specificity.

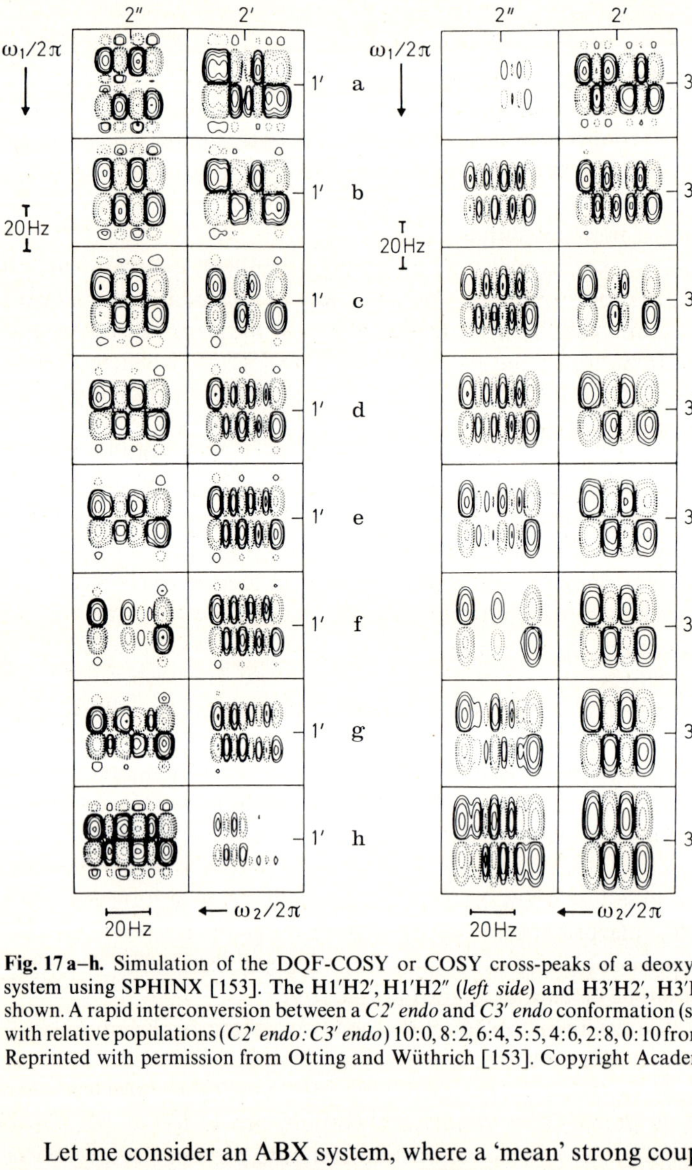

Fig. 17 a–h. Simulation of the DQF-COSY or COSY cross-peaks of a deoxyribose proton spin-system using SPHINX [153]. The H1′H2′, H1′H2″ (*left side*) and H3′H2′, H3′H2″ cross-peaks are shown. A rapid interconversion between a *C2′ endo* and *C3′ endo* conformation (see Fig. **e**) is assumed with relative populations (*C2′ endo*: *C3′ endo*) 10:0, 8:2, 6:4, 5:5, 4:6, 2:8, 0:10 from [*A*] down to [*H*]. Reprinted with permission from Otting and Wüthrich [153]. Copyright Academic Press Inc (1987)

Let me consider an ABX system, where a 'mean' strong coupling parameter, θ can defined (see above). Keller et al. [156] have shown that the *initial build-up rate* of the AX cross-peak (which is a reliable estimate of the AX distance in other cases) is now given by:

$$\cos^2\theta\,\sigma_{AX} + \sin^2\theta\,\sigma_{BX}$$

Inferring structure from such a build-up rate become hazardous, since, even if multi-spin effects have been avoided by choosing a sufficiently short mixing period, a third spin still interferes. Furthermore, the relative intensities of the multiplet components are also modified. The spectroscopist might be deceived be these effects if she (or he) intends to use the fine structure as an assignment aid. As long as one deals with moderately strong coupling, it is still possible to derive quantitative N.O.E., build-up rate by using a read pulse of a small angle (10°–20°) [156] at the expense of the sensitivity.

As mentioned earlier, strong couplings are frequently encountered in proteins. Overhauser effects involving geminal pairs are not only biased by strong coupling but also by spin diffusion as a result of the extraordinarily short H–H distance (1.78 Å). From their analysis of glycine spin systems is NOESY, Kay et al [158] concluded that spin diffusion remains the major concern for practical mixing times. Moreover, disregarding strong coupling is probably not of great consequence since, in general, it is not possible to integrate accurately cross-peaks in extreme cases ($J/\delta > 0.5$).

In conclusion, crude guidelines can be provided for analyzing cross-peaks involving strongly coupled pairs in NOESY. If one deals with geminal proton pairs (CH_2 groups in proteins or DNA), there is every chance that for realistic mixing times (say > 50 ms) errors due to multispin effects will dominate. In other cases (for instance, protons of a sugar ring) no clear-cut conclusion can be drawn, in view of the less efficient spin diffusion due to larger H–H distances.

8 Heteronuclear Chemical Shift Correlation of Biomolecules

A wide range of 2D NMR experiments, sketched in Fig. 18, is available for studies of scalar connectivities between ^{1}H and heteronuclei present in biomolecules: ^{13}C and ^{15}N in proteins [46, 159], ^{13}C and ^{31}P in oligonucleotides [64–66] and ^{13}C in oligosaccharides. When rare spin 1/2 nuclei are bound covalently to a biopolymer, valuable information can be gathered, as for example on the geometry of the metal-sulfur cluster of metallothionein-2 by ^{113}Cd–^{1}H correlation [160]. So far, heteronuclear experiments have been of real help for resonance assignments, but they have the potential to provide structural information when the signal-to-noise ratio is high enough for this purpose. Carbohydrates, the ^{1}H NMR spectra of which are notoriously difficult to unravel, illustrate the usefulness of ^{13}C–^{1}H correlation: the assignment of the nonanomeric protons is made easier [161, 162] by the greater spread in ^{13}C chemical shifts [163].

The same correlation experiment can be recorded in many different ways, because either nucleus can be detected, pulses can be selectively applied to either nucleus and lastly heteronuclear decoupling can be employed at will. Depending on the hardware and software capabilities of the spectrometer, one can choose among several options which differ in their relative sensitivity. In fact, an experiment can start with either ^{1}H or X polarization (named spin A for convenience), and can end with the detection of either ^{1}H or X magnetization (spin B). If the repetition rate limitation is disregarded (approximation of short

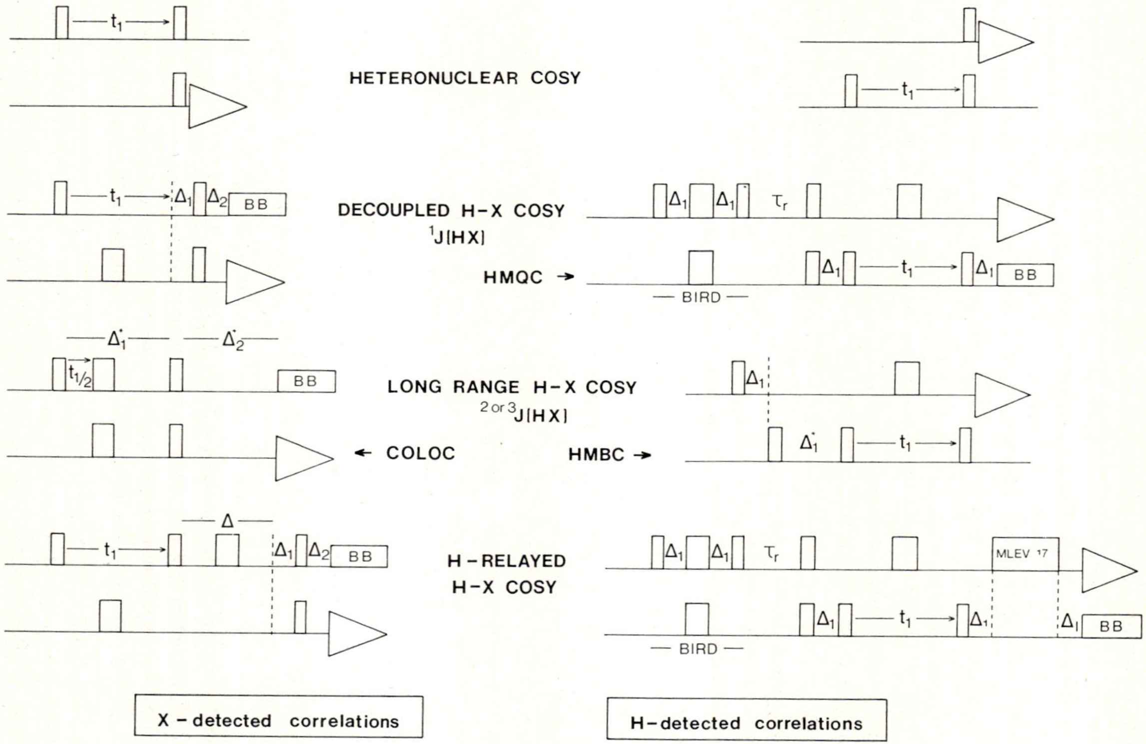

Fig. 18. Various pulse schemes used for heteronuclear 2D NMR of biomolecules (X stands for ^{13}C, ^{15}N or ^{31}P). The delays Δ_1 and Δ_2 are turned for one-bond J_{XH} (typically 70 to 140 Hz) and Δ_1^* and Δ_2^* for long-range J_{HX} (typically 2 to 10 Hz)

relaxation times), the sensitivity of the correlation experiment is proportional to $\gamma_A \cdot \gamma_B^{3/2}$ [47]. Thus maximal sensitivity is obtained if the experiment starts *and* ends on protons ($A = B = {}^1H$), but this design requires ^1H observation while applying X-pulses, a hardware requirement only recently made available as a standard feature of commercial spectrometers.

As a consequence of the low natural abundance of ^{13}C and ^{15}N, ^1H detected versions require excellent spectrometer stability, since the cancellation of the 100-fold larger unwanted signals of the $^{12}C{-}^1H$ or $^{14}N{-}^1H$ pairs is mainly accomplished by means of phase cycling after two scans. In addition to this severe dynamic range problem, one may have to cope with solvent suppression [148] if exchangeable protons are being investigated. Therefore, for biopolymers in H_2O, one may be tempted to resort to the easy but far less sensitive solution of choosing X detection. Yet, ^1H detection is by far the superior choice for many biomolecules because of limited sample quantities, even if more effort has to be devoted to careful experimental setup.

Let us now briefly discuss the various pulse schemes for heteronuclear J-correlation. The two experiments sketched at the top of Fig. 18 are heteronuclear versions of COSY [43, 164]. The absence of diagonal peaks in heteronuclear correlation removes the major limitation of homonuclear COSY: i.e., the dispersive diagonal. The cross-peaks exhibit the same 2D antiphase multiplet structure, from which the J-coupling can be estimated for structural studies. Either ^1H- or X- detected versions show quite similar sensitivity (the experiment starts on one spin and is detected *via* the other one). This simple experiment is suitable for easy cases, where sensitivity requirements are not important and where all J's are of a similar order of magnitude. These conditions are fulfilled in oligonucleotides, in which the 100% abundant ^{31}P nuclei display several $^3J_{PH}$ and $^4J_{PH}$ [65] but no ^1J. The ^{31}P and ^1H resonances of a dodecamer have been assigned sequentially by means of this simple experiment by Sklenar et al. (Fig. 19) [67]. ^1H detection is more suitable in this case because high digital resolution in the proton dimension is essential for unravelling the very crowded $H_{3'}$, $H_{4'}$ and $H_{5'}$ region (4 protons per ribose in less than 1 ppm).

Although coupling constants gleaned from an intact multiplet structure contain useful structural information, recording multiplets has a major drawback, because the intensity is divided among more lines. Therefore when one aims only at assigning either spectrum, it is desirable to remove the heteronuclear coupling so that a connectivity gives rise to a singlet in both dimensions (the $^1H{-}^1H$ couplings remain). In the X-detected version, decoupling is achieved by a 180° pulse during t_1 and by standard broadband decoupling during t_2 [165, 166]. The requirement for complete decoupling over the entire ^1H chemical shift range suggests the use of composite pulse decoupling [167] rather than standard noise decoupling in order to minimize the r.f. power in the probe (this will be a major concern for spectrometers at higher fields). The delays Δ_1 and Δ_2 (see Fig. 18), required for interconverting antiphase coherence into in-phase coherence and vice versa, must be tuned for $^1J_{XH}$, the value of which is weakly dependent on the nature of the adjacent chemical bonds. The ^1H detected version

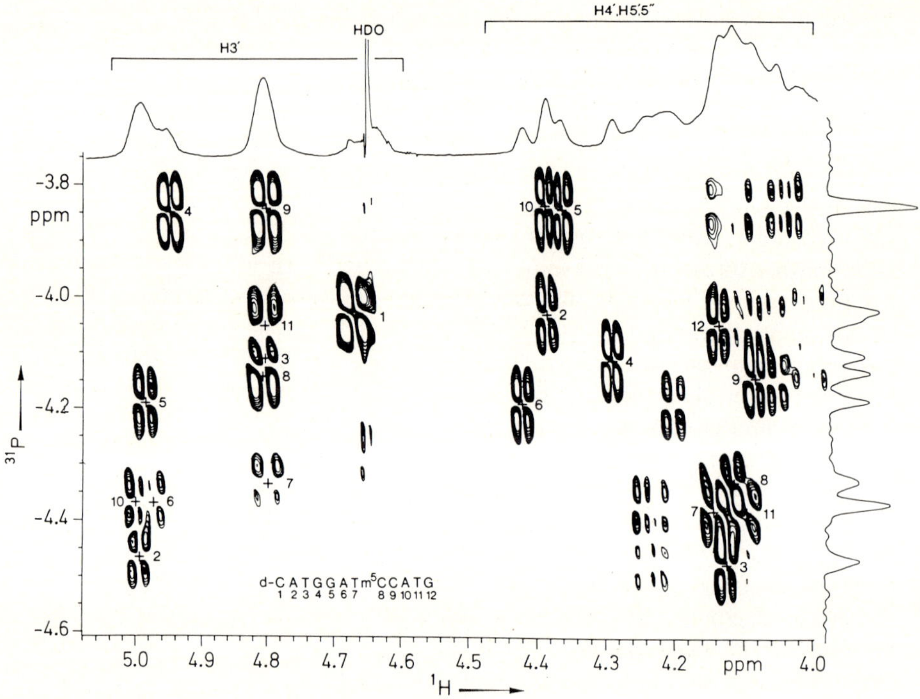

Fig. 19. 2D absorption mode ^{1}H–^{31}P correlation spectrum of the dodecamer d(CATGGATm^{5}CCATG) recorded using ^{1}H-detected heteronuclear COSY (see Fig. 18). Protons resonances were presaturated during the relaxation delay using a series of 180° pulses. 600 A_{260} units were dissolved in 0.4 ml D_2O, 0.1 NaCl and 10 mM sodium phosphate p $^{2}H = 7.4$. The total measuring time is 8 h. Reprinted with permission from Sklenář et al. (67) Copyright Fed of Eur Biochemical Societies, Elsevier Science Publisher B.V. (1987)

of heteronuclear correlation, called *heteronuclear multiple quantum correlation* (HMQC) [168, 169], is more sensitive owing to the fact that ^{1}H magnetizations are involved at the beginning and at the end of the sequence. To diminish the problem of the unwanted ^{12}C–^{1}H signals, this sequence can be improved by the addition of a BIRD sequence [170]. At the beginning of the experiment, the corresponding magnetizations vanish so that residual signals can be removed more efficiently by phase cycling.

Recent advances in ^{13}C–^{1}H correlation experiments have improved their *sensitivity* beyond a spectroscopist's wildest dreams. Early attempts on proteins relied on ^{13}C detection and required very large sample quantities and long experimental times or isotope enrichment [171, 172]. In 1982, Chan and Markley [171] reported the ^{13}C–^{1}H correlation map of a ferredoxin (MW = 11 000). With 20% ^{13}C enrichment and 150 mg of material in a 20 mm O.D. tube, a spectrum was obtained in 50 hours. In 1987, Sklenar and Bax [48] demonstrated the use of indirect detection on hen egg white lysozyme (Fig. 20) (MW = 14 000). Data were

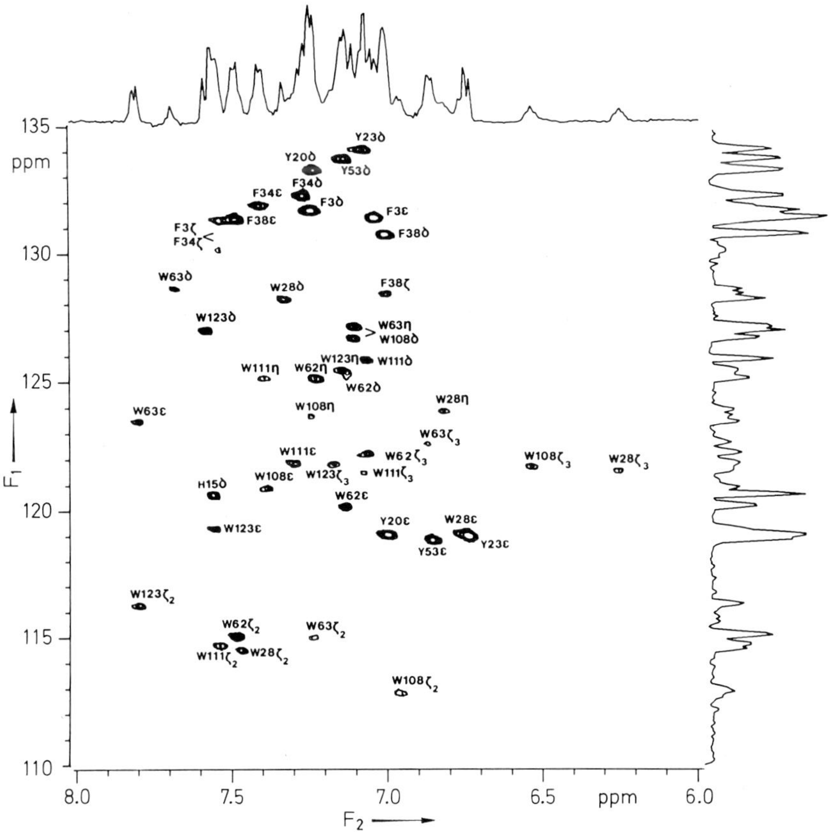

Fig. 20. 2D absorption mode heteronuclear shift correlation (through ^1J) of the aromatic residue in hen egg white lysozyme (see text for details). Reprinted with permission from Sklenar and Bax [48]. Copyright Academic Press Inc (1987)

collected in 20 hours on a 40 mg sample without any enrichment in a 5 mm O.D. tube. Comparison of these two results (obtained at nearly the same field strength) is evidence of the major contribution that pulse sequence can make for better results at high fields.

So far, we have dealt with systems in which the largest coupling constants were observed ($^1J_{CH}$ or $^1J_{NH}$ in proteins or oligosaccharides). If either the ^1H or the X spectrum has not been already nearly completely assigned by other means, this type of correlation is not of great help for assignment. An X resonance is only associated with the resonance of the attached proton, but nothing more. So, it might be useful to have at one's disposal additional tools capable of correlating X nuclei with other protons using either long-range J_{XH} or a combination of $^1J_{XH}$ and proton-to-proton relay.

Because of the antiphase nature of J-correlation cross peaks, larger J-couplings usually give rise to larger correlation peaks than do smaller couplings. Thus, the use of *heteronuclear long range correlation* for protonated carbons or nitrogens requires the suppression of peaks resulting from the directly attached protons. In the X-detected version (named COLOC [62, 173, 174]), a compromise value of the delay Δ_1^* has to be chosen to yield both adequate suppression of the $^1J_{XH}$ and efficient excitation of the long range. However, regardless of the choice of Δ_1^*, the spectral intensities will still be modulated by $^1J_{XH}$ and by homonuclear H–H couplings. Kessler and coworkers [61, 62, 175] and others [176] have made extensive use of the COLOC technique for assigning ^{13}C and ^1H resonances in peptides and glycopeptides (see Fig. 5). In peptides, each carbonyl resonance exhibits a cross-peak with at least three protons: the $C^{\alpha}H$ of the same residue (^2J), the NH (^2J) and the $C^{\alpha}H$ (^3J) of the following residue. Additional connectivities with the $C^{\beta}H_2$ of the same residue provide not only the *stereospecific assignment* (in conjunction with ^1H spectral parameters) but also acute insight into both the conformation and the dynamics of the side-chain. In this context, the use of high-field spectrometer is crucial for two reasons: first for compensating the intrinsically weak sensitivity of the experiment (cross-peak cancellation due to line broadening); and second for increasing the separation within $C^{\beta}H$ pairs. Even partial degeneracy in these pairs is an insurmountable obstacle in structural analysis of the side-chain.

A substantial gain in sensitivity can be achieved by ^1H detection of long range heteronuclear correlation [177]. The HMBC experiment (*Heteronuclear Multiple Bond Correlation* (Fig. 18)) is an adaptation of the HMQC experiment used for one-bond correlation. Here, spectrometer stability is a more pressing requirement because the unwanted signals from protons bound to ^{12}C can be suppressed only by phase cycling, for want of an efficient BIRD sequence for small couplings. One aspect of the stability problem is major fluctuations that are supposed to be compensated by both the ^2H lock and, if available, an automatic shim device. The efficiency of both relies not only on their correct setting (level and time constant) but also on the spectral characteristics of the ^2H signal (line width and possible multiplet structure), upon which the spectrometer is locked. Cross talk between the various channels (^1H, ^2H and X) may also perturb the lock signal. Provided that these settings are optimized, the HMBC experiment is preferable to the COLOC owing to its greater sensitivity. In small oligosaccharides such as α-Neu5Ac–(2–3)βGal–(1–4)Glc, shown in Fig. 3: interglycosidic connectivities between Ney5Ac C-2 and Gal H-3 and between Gal C-1 and Glc H-4 could be detected in a overnight experiment on 1.2 mg of the α-anomer (see Fig. 21) [162]. This could be used to sequence branched oliogosaccharides which are difficult to sequence by other means.

Fig. 21 a, b. ^1H–^{13}C chemical shift correlation (long range connectivities) in αNeu5Ac (2–3) βGal-(1–4)Glc, a trisaccharide, the covalent structure of which is given in the lower half of Fig. 3. The data were recorded in 16 h using the HMBC experiment (see Fig. 18). **b)** is an expansion of the most crowded spectral region of **a)**. Several interresidue connectivities used for the assignment are labelled (GalH3 N2), (GlcH4 G1) and (GalH1 B4) (see text) Reprinted with permission from Lerner and Bax [162]. Copyright Elsevier Science B.V. (1987)

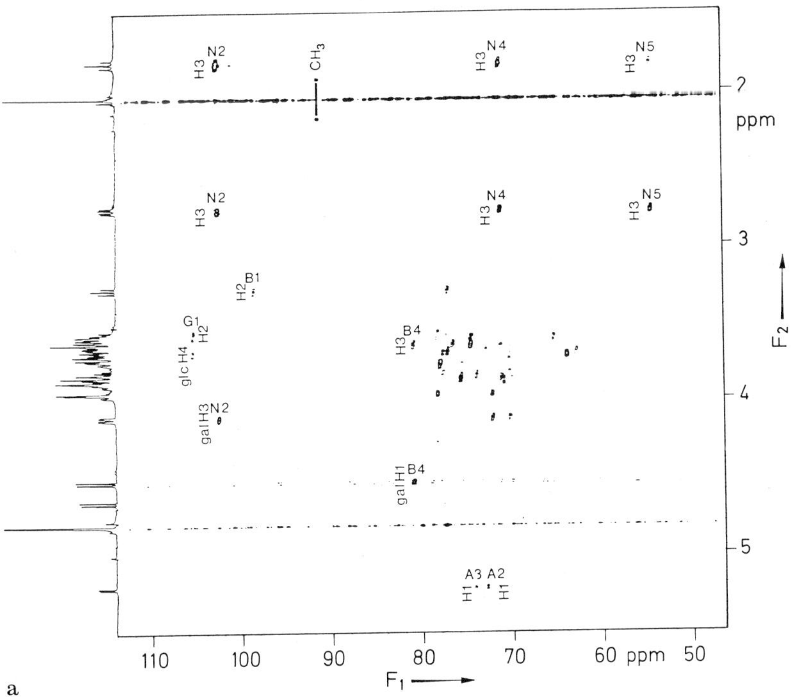

H3 N2
CH_3
H3 N4
H3 N5
ppm
H2 B1
G1 H2
glc H4
H3 B4
gal H3 N2
gal H1 B4
A3 A2
H1 H1
110
100
90
80
70
60 ppm 50
F_1
F_2

a

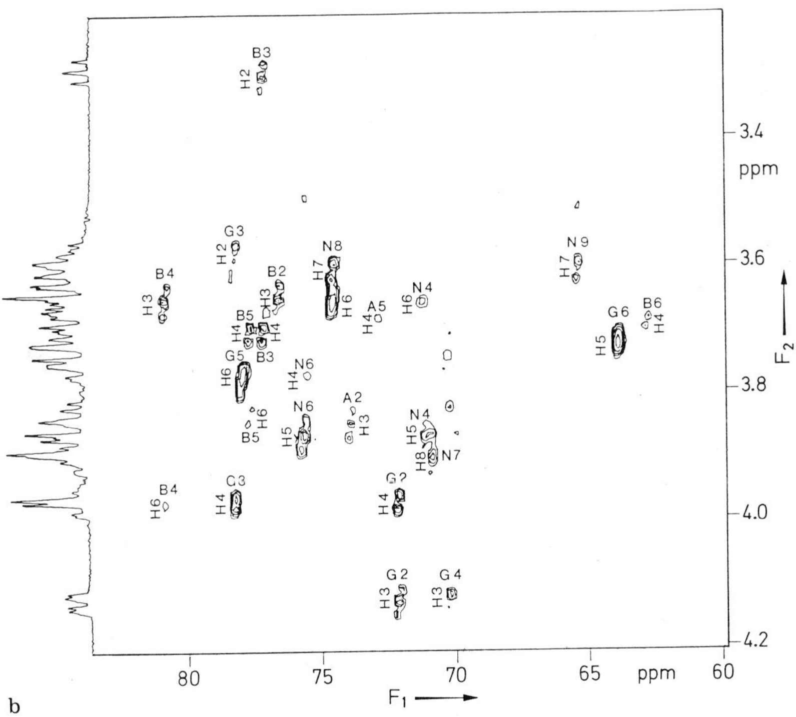

B3
H2
3.4
ppm
G3
H2
N8
H7
N9
H7
B4
H3
B2
H3
N4
H6
A5
H4
H6
G6
B6
H4
H5
B5
H4
H4
G5
B3
N6
H4
H6
H6
B5
N6
H5
A2
H3
N4
H5
H8
N7
B4
H6
G3
H4
G2
H4
G2
H3
G4
H3
3.6
3.8
4.0
4.2
80
75
70
65
ppm
60
F_1
F_2

b

When the coupling are too small to be detected by a long-range correlation experiment (such bad luck originates either from the macromolecule's conformation or its slow tumbling) an alternate method can be tried, based on a combination of a X–H transfer and either a one-step [178, 179] or a Hartman–Hahn relay [180]. Let me point out that on large macromolecules with slow motions, it is likely that the relay experiment will also fail. On small peptides [179] or oligosaccharides [162], the heteronuclear relayed correlation is of great help only for assignment: no heteronuclear ^3J is involved here and thus no conformational information, additional to that more easily obtained by ^1H, will be gained.

9 Conclusion and Perspectives

In the previous sections, the methods available for structural studies of biomolecules of molecular weight less than 8000–12000 have been discussed. Without any chemical or biological modification of a protein, oligonucleotide or carbohydrate, resonance assignment can be achieved. Spectral parameters, converted into conformational ones, can be used as input for various refinement programs for building the macromolecule's conformation [181–183].

At the field strengths currently available, resonance overlap is a major limitation for tackling larger molecular weights, in addition to line broadening which decreases sensitivity dramatically. In an unpredictable manner, some macromolecules may be, more NMR-friendly than others, because residual internal freedom restores sufficiently narrow line-width. Since internal flexibility also means conformational averaging, the line-width advantage will be paid at the expense of the quality of the conformational data, which may or may not permit derivation of a reliable preferred conformation. Random fractional deuteration [131] has been mentioned already as a solution to line broadening, but labeling techniques are required. Hence, this approach excludes biopolymers which cannot synthesized, cloned into microorganisms or artificially grown in high yield in cell culture. With the same restriction, ^1H spectra of proteins and other macromolecules can be simplified by ^{13}C and ^{15}N labeling. As a result of residue selective labeling, protons directly bound to ^{13}C or to ^{15}N (and thus already identified as belonging to a give type of residue) can be singled out on the basis of the large heteronuclear $^1J_{XH}$ [184–188]. This coupling is much larger than the J_{HH} and thus the cancellation due to broad antiphase line is avoided. Consequently, these filtration techniques can be applied to larger macromolecules than can experiments which utilize the J_{HH}. Early filtering experiments were obtained by the difference of 1D recording with and without broadband decoupling on the X nuclei [184–186]. Similarly any 2D proton experiment can be modified to incorporate an X-filter, which will act differently on protons, depending on whether or not they bear a ^{13}C or a ^{15}N. During even-numbered scans all ^1H magnetizations, regardless of their X-partners, are in phase while

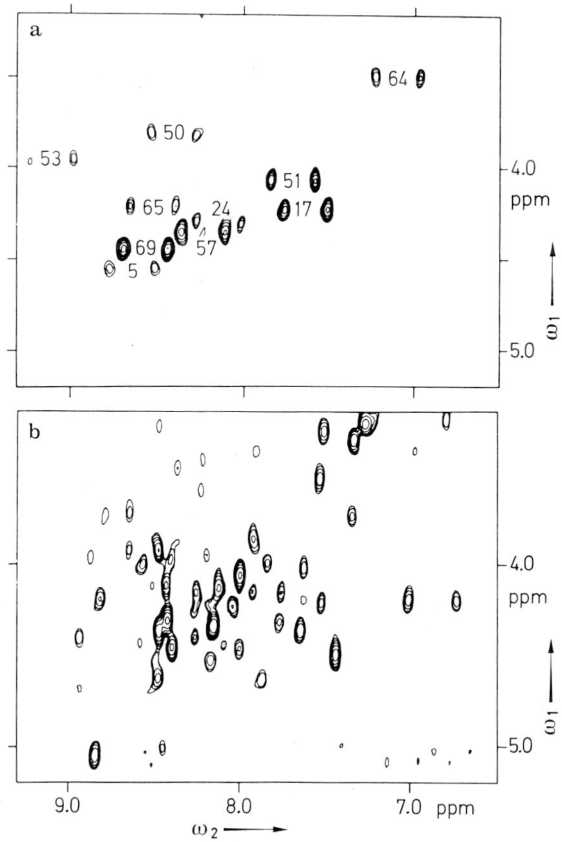

Fig. 22 a, b. ^1H TOCSY (or HOHAHA) with ^{15}N filtering of the DNA-binding domain 1–76 of c2 repressor from *Salmonella* phage P22 with all 10 leucyl residue labelled with ^{15}N in the extent > 85%. Protein concentration 8 mM, solvent H_2O. **a**) the NH–$C^{\alpha}H$ cross-peaks of the Leu residues are indentified with sequence location. **b**) non-labeled amino acids. Reprinted with permission from Otting et al. [187]. Copyright Academic Press Inc (1986)

during odd-numbered scans, magnetizations of ^{13}C or ^{15}N attached protons exhibit a phase opposite to all other magnetizations. Coaddition or cosubstraction of the two data sets selects either the labelled or the unlabelled resonances [187–189] (see Fig. 22). These techniques, recently reviewed by Griffey and Redfield [190] are very promising for larger macromolecules.

A very commonly used procedure for reducing the ambiguities related to spectral overlaps relies on the fact that the chemical shifts of many protons show different pH and temperature dependences. In this respect, this tedious technique, actually limited by the stability range of the macromolecule, is nothing else than the introduction of a third dimension: a more elegant approach utilizes the extension of 2D NMR to three-dimensional NMR [197] (3D-NMR). A 3D-NMR experiment can be conceived as the combination of two complementary

2D experiments, for instance a NOESY with an HOHAHA [200, 201]. In this example, the magnetization first carried by an NH is transferred by J-coupling to the $C^{\alpha}H$ of the same residue and then to the NH of the next residue via N.O.E.: a cross-peak is thus observed in a 3D space at the chemical shift of the involved nuclei and the occurrence of overlaps thus decreases by a fair amount. However, the sensitivity of this experiment is limited by the efficiency of the N.O.E. and J-transfer: cross-peaks in homonuclear 2D experiments are usually at most 10% of the diagonal, and, as a result of the double transfer, 3D homonuclear cross-peaks are hundred-fold smaller than the corresponding diagonal. This dynamic range problem, added to the suppression of the solvent noted earlier severely decreases the sensitivity of the 3D homonuclear techniques.

Homonuclear 2D experiments can also be combined with 2D heteronuclear ones for instance an NOESY with an HMQC sequence. These heteronuclear 3D techniques [202–204] can be much more sensitive (provided that isotopic enrichment is used) since the heteronuclear transfer *via* the one-bond J-coupling can be oprimized to nearly 100%. In these types of experiments (NOESY-HMQC or HOHAHA-HMQC) the number of resonances observed is similar to 2D spectra, and the assignment strategy described above can be used with only minor changes. The application of these promising techniques to Staphylococcal nuclease, the anaphylatoxin C5a, interleukin 1β and calmodulin (a Ca-binding protein) clearly demonstrated their potential power for spectral simplification, as they generate a series of 2D NOESY spectra edited by the chemical shift of the directly bonded ^{15}N atoms. Without any increase of the magnetic field, the study of 150 to 200 residue proteins becomes feasible in the next future by splitting the 2D map into several ones (according either to a specific label in 2D experiment or the ^{15}N chemical shift in 3D experiments) under two stringent conditions. First, an acceptable line-width is mandatory and second, isotope labeling technique should be feasible for the macromolecule at a reasonable cost.

Whereas a decisive increase of the available magnetic field would obviously make conformational studies easier, other improvements in NMR methods are likely to be available sooner. Until recently, the Cooley-Tukey Fast Fourier Transform (FFT) algorithm [191] has been the only efficient way for obtaining a frequency domain spectrum—on which the interpretation is done—from the time domain free induction decay (FID) actually sampled. Ernst [192] proposed that more information could be obtained from the spectrum by multiplying the FID by a suitable apodizing function prior to the FFT but there is always a trade-off between noise suppression and resolution enhancement. Linear prediction (LP) [193–194] and maximum entropy (MEM) [195, 195] methods have been proposed as a better alternative for alleviating this compromise. LP tries to fit damped sinusoids to the acquired FID. MEM aims at reconstructing the most likely spectrum by maximizing its entropy. According to early users of MEM [196] "*for a given measuring time, a 2D NMR spectrum with higher resolution in* ω_1 *can be obtained; instead the same resolution could be obtained with a shorter measuring time*". Both methods, which in fact require much longer computing time than FFT, could not be considered several years ago, but should now benefit from the increased power and speed of the new generation of computers equipped with array processors.

In the course of this review, it has been emphasized that high fields have been the major cause of success in structural studies of biopolymers in solution. However, the arsenal of the 2D pulse (and now 3D) sequences at the spectroscopist's disposal for selecting and enhancing spectral information has played the second major role and it is likely that, without it, most of the potential of high fields would have been wasted. In 1976, Ernst and coworkers [81] claimed in their pioneer work on 2D NMR that "*2D spectroscopy fascinates by its conceptual simplicity and by its general applicability. It seems to open a further dimension to the spectroscopist.*" Let us believe them again for the next decade and ... for the next dimension.

Acknowledgments: The author wishes to thank Dr. Ad Bax, Prof. Dr. Kurt Wüthrich, Prof. Robert Kaptein and Prof. Dr. Horst Kessler for their permission to reproduce published material, Drs. Ad Bax and Laura Lerner for helpful discussions and comments. This work was supported by the *Centre National de la Recherche Scientifique* and by the C.N.R.S.—N.I.H. international exchange agreement.

References

1. Williams TC, Shelling JG, Sykes BD (1986) NATO ASI Ser., Ser A107 (NMR Life Sci): 93
2. Detellier C, Laszlo P (1985) Proc Indian Acad Sci 94: 291
3. Davis GD, Tosteson DC (1975) Biochemistry 14: 3962
4. Vishwanath CK, Easwaran KRK (1982) Biochemistry 21: 2612
5. Brévard C, Granger P (1981) Handbook of High Resolution Multinuclear NMR, Wiley and sons, New York
6. Urry DW, Trapane TL, Venkatachalan CM, Prasad KU (1983) J Phys Chem 87: 2918
7. Svard M, Drakenberg T (1986) Acta Chem Scand B40: 689
8. Hinton JF, Turner GL (1984) J Magn Reson 59: 262
9. Patel DJ (1974) Biochemistry 13: 2388
10. Reid DG, Salisbury SA, Williams DH (1983) Biochemistry 22: 1377
11. Gaugain B, Markovits J, Le Pecq J-B, Roques BP (1981) Biochemistry 20: 3035
12. Patel DJ (1979) Eur J Biochem 99: 369
13. Klevit RE, Wemmer DE, Reid BR (1986) Biochemistry 25: 3296
14. Neidle S, Pearl LH, Skelly JV (1987) Biochem J 243: 1
15. Nuss ME, James TL, Apple MA, Kollman PA (1980) Biochim Biophys Acta 609: 136
16. Patel DJ, Shapiro L (1986) J Biol Chem 261: 1230
17. Brown SC, Mullis K, Levenson C, Shafer RH (1984) Biochemistry 23: 403
18. Quinn AJ, Chapman D (1980) CRC Crit Rev Biochem 8: 1
19. Griffin RG (1981) Meth Enzymol 72: 108; Smith ICP, Ekiel IH (1984) In: Gorenstein DG (ed) Phosphorus-31 NMR: Principles and Application, pp 447–475 Academic, New York; Jacobs RE, Oldfield E (1981) Prog NMR Spectros 14: 113; McDonald PM, Sykes BD, McElhaney RN (1984) Can J Biochem Cell Biol 62: 1134; Davis JH (1986) Chem Phys Lipids 40: 223
20. Seelig J, Tamm L, Hymel L and Fleischer S (1981) Biochemistry 20: 3922
21. Seelig J, McDonald PM (1987) Acc Chem Res 20: 221
22. Campbell RF, Meirovitch E, Freed JH (1979) J Phys Chem 83: 525

23. Davis JH, Jeffrey KR, Bloom M, Valic MI, Higgs TP (1976) Chem Phys Lett 42: 390
24. Rance M, Byrd RA (1983) J Magn Reson 52: 221
25. Seelig J, Borle F, Cross TA (1985) Biochim Biophys Acta 814: 195
26. Speyer JB, Sripada PK, Das Gupta SK, Shipley GG, Griffin RG (1987) Biophys J 51: 687
27. Felsenfeld G (1985) Sci Am 253: 58
28. Doolittle RF (1985) Sci Am 253: 88
29. Sharon N (1980) Sci Am 243: 90
30 Kennedy JF, White CA (1983) Bioactive Carbohydrates in Chemistry, Biochemistry and Biology, Ellis Horwood, Chichester (UK)
31. Ramachadran GN, Sasisekharan V (1968) Adv Protein Chem 23: 283
32. Schulz GE, Schirmer RH (1979) Principle of Protein Structure, Springer, Berlin Heidelberg New York
33. Walsh KA, Ericsson LH, Parmelee DC, Titani K (1981) Ann Rev Biochem 50: 261
34 Otting G, Marchot P, Bougis PE, Rochat H, Wüthrich K (1987) Eur J Biochem 168: 303
35. Wüthrich K (1986) NMR of Proteins and Nucleic Acids, Wiley, New York
36. Dickerson RE (1983) Sci Am 249: 94
37. Drew H, Takano T, Tanaka S, Itakura K, Dickerson RE (1980) Nature 286: 567
38. Wang AH-J, Quigley GJ, Kolpak FJ, Crawford JL, van Boom JH, van der Marel G, Rich A (1979) Nature 282: 680
39. Altona C (1982) Rec Trav Chim Pays-Bas 101:413; Rinkel LJ, Altona C (1987) J Biomol Struct Dyn 4: 621
40. Saenger W (1984) Principles of Nucleic Acid Structure, Springer, Berlin Heidelberg New York
41. Koerner TAW, Prestegard JH, Yu RK (1987) Meth Enzym 138: 38
42. von Philipsborn W, Mueller R (1986) Angew Chem Int Eng Ed 25: 383
43. Maudsley AA, Ernst RR (1977) Chem Phys Lett 50: 368
44. Maudsley AA, Müller L, Ernst RR (1977) J Magn Reson 28: 463
45. Bax A, Griffey RH, Hawkins BL (1983) J Magn Reson 55: 301
46. Live DH, Davis DG, Agosta WC, Cowburn D (1984) J Amer Chem Soc 106: 6104
47. Ernst RR, Bodenhausen RR, Wokaun A (1987) Principles of Nuclear Magnetic Resonance in One and Two Dimensions, Clarendon, Oxford
48. Sklenar V, Bax A (1987) J Magn Reson 71: 379
49. Bolton PH, Mirau PA, Shafer RH, James TL (1981) Biopolymers 20: 435
50. Neumann JM, Tran-Dinh S (1981) Biopolymers 20: 89
51. Gorenstein DG, Luxon BA (1979) Biochemistry 18: 3796
52. Werbelow LG, Grant DM (1977) Adv Magn Reson 9: 189
53. Vold RL, Vold RR (1978) Prog NMR Spectrosc 12: 79
54. Lenk R (1977) Brownian Motion and Spin Relaxation, Elsevier, New York
55. Richarz R, Nagayama K, Wüthrich K (1980) Biochemistry 19: 5189
56. Henry GD, Weiner JH, Sykes BD (1986) Biochemistry 25: 590
57. Bystrov VF (1976) Prog NMR Spectrosc 10: 41
58. Saunders M, Wishnia A, Kirkwood JG (1957) J Amer Chem Soc 79: 3289
59. Garbay-Jaureguiberry C, Marion D, Fellion E, Roques BP (1982) Int J Pept Protein Res 20: 342
60. Roques BP, Rao R, Marion D (1980) Biochimie 62: 753
61. Kessler H, Griesinger C, Wagner K (1987) J Am Chem Soc 109: 6937
62. Kessler H, Griesinger C, Zarbock J, Loosli HR (1987) J Magn Reson 57: 331
63. Kessler H, Bermel W, Griesinger C, Kolar C (1986) Angew Chem Int Ed Eng 25: 342
64. Pardi A, Walker R, Rapoport H, Wider G, Wüthrich K (1983) J Amer Chem Soc 105: 1652
65. Marion D, Lancelot G (1984) Biochem Biophys Res Commun 124: 774
66. Lai K, Shah DO, DeRose E, Gorenstein DG (1984) Biochem Biophys Res Commun 121: 1984
67. Sklenář V, Miyashiro H, Zon G, Miles HT, Bax A (1987) FEBS Lett 208: 94
68. Billeter M, Braun W, Wüthrich K (1982) J Mol Biol 55: 321
69. Wüthrich K, Billeter M, Braun W (1983) J Mol Biol 69: 949
70. Clore GM, Gronenborn AM (1985) FEBS Lett 179: 187
71. Englander SW, Wand AJ (1987) Biochemistry 26: 5953
72. Di Stefano DL, Wand AJ (1987) Biochemistry 26: 7272
73. Kemmink J, Boelens R, Koning TMG, Kaptein R, van der Marel GA, van Boom JH (1987) Eur J Biochem 162: 37
74. Tomic MT, Wemmer DE, Kim S-H (1987) Science 238: 1722
75. Clore GM, Sukumaran DK, Nilges M, Gronenborn AM (1987) Biochemistry 26: 1732; Zarbock J, Clore GM, Gronenborn AM (1986) Proc Natl Acad Sci USA 83: 7628

76. Wand AJ, Di Stefano DL, Feng Y, Roder H, Englander SW (1989) Biochemistry 28: 186
77. Aue WP, Karhan J, Ernst RR (1976) J Chem Phys 64: 4226
78. Nagayama K, Wüthrich K, Bachmann P, Ernst RR (1977) Biochem Biophys Res Commun 78: 99
79. Nagayama K, Bachmann P, Wüthrich K, Ernst RR (1978) J Magn Reson 31: 133
80. Marion D, Wüthrich K (1983) Biochem Biophys Res Commun 113: 967
81. Aue WP, Bartholdi E, Ernst RR (1976) J Chem Phys 64: 2229
82. Nagayama K, Wüthrich K, Ernst RR (1979) Biochem Biophys Res Commun 90: 305
83. Jeener J, Meier BH, Bachmann P, Ernst RR (1979) J Chem Phys 71: 4546
84. Macura S, Ernst RR (1980) Mol Phys 41: 95
85. Bodenhausen G, Kogler H, Ernst RR (1984) J Magn Reson 58: 370
86. Eich GW, Bodenhausen G, Ernst RR (1982) J Am Chem Soc 104: 3731
87. Braunschweiler L, Ernst RR (1983) J Magn Reson 53: 521
88. Wokaun A, Ernst RR (1978) Mol Phys 40: 1153
89. Sørensen OW, Eich GW, Levitt MH, Bodenhausen G, Ernst RR (1983) Prog NMR Spectrosc 16: 163
90. van de Ven FJM, Hilbers CW (1983) J Magn Reson 54: 512
91. Packler KJ, Wright KM (1983) Mol Phys 50: 797
92. Widmer H, Wüthrich K (1987) J Magn Reson 74: 316
93. Kay LE, Scarsdale JN, Hare DR, Prestegard JH (1986) J Magn Reson 68: 515
94. Piantini U, Sørensen OW, Ernst RR (1982) J Amer Chem Soc 104: 6800
95. Bax A, Freeman R (1981) J Magn Reson 44: 542
96. Rance M, Wagner G, Sørensen OW, Wüthrich K, Ernst RR (1984) J Magn Reson 59: 250
97. Rance M, Sørensen OW, Bodenhausen G, Wagner G, Ernst RR, Wüthrich K (1983) Biochem Biophys Res Commun 117: 479
98. Bernstein MA, Hall LD (1982) J Amer Chem Soc 104: 5553
99. Gasa S, Nakamura M, Makita A, Ikura M, Hikichi K (1986) Eur J Biochem 155: 603
100. Neuhaus D, Wagner G, Vasak M, Kägi JHR, Wüthrich K (1985) Eur J Biochem 151: 257
101. Bax A, Drobny G (1985) J Magn Reson 61: 306
102. Wagner G (1983) J Magn Reson 55: 151
103. Chazin WJ, Wüthrich K (1987) J Magn Reson 72: 358
104. Homans SW, Dwek RA, Fernandes DL, Rademacher TW (1984) Proc Natl Acad Sci USA 81: 6286
105. Sørensen OW (1984) Dissertation ETH Zürich
106. Weber PL, Müller L (1987) J Magn Reson 73: 184
107. Bax A. Davis DG (1985) J Magn Reson 65: 355
108. Homans SW, Dwek RA, Boyd J, Soffe N, Rademacher TW (1987) Proc Natl Acad Sci USA 84: 1202
109. Braunschweiler L, Bodenhausen G, Ernst RR (1983) Mol Phys 48: 535
110. Otting G, Wüthrich K (1986) J Magn Reson 66: 359
111. Sørensen OW, Levitt MH, Ernst RR (1983) J Magn Reson 55: 104 and Rance M, Sørensen OW, Leupin H, Kogler H, Wüthrich K, Ernst RR (1985) J Magn Reson 61: 67
112. Dalvit C, Wright PE (1987) J Mol Biol 194: 313
113. Dalvit C, Wright PE, Rance M (1987) J Magn Reson 71: 539
114. Williamson MP, Marion D, Wüthrich K (1984) J Mol Biol 173: 341
115. Otting G, Steinmetz WE, Bougis PE, Rochat H, Wüthrich K (1987) Eur J Biochem 168: 609
116. Wörgötter E, Wagner G, Vasak M, Kägi JHR, Wüthrich K (1987) Eur J Biochem 167: 457
117. Kline AD, Wüthrich K (1986) J Mol Biol 192: 869
118. van der Ven FLM, Hilbers CW (1986) J Mol Biol 192: 419
119. Solomon I (1955) Phys Rev 99: 559
120. Noggle JH, Schirmer RE (1971) The nuclear Overhauser effect: chemical applications, Academic, New York
121. Kalk A, Berendsen HJC (1976) J Magn Reson 24: 343
122. Anil-Kumar, Ernst RR, Wüthrich K (1980) Biochem Biophys Res Commun 95: 1
123. Marion D, Genest M, Caille A, Peypoux F, Michel G, Ptak M (1986) Biopolymers 25: 153
124. Macura S, Huang Y, Suter D, Ernst RR (1981) J Magn Reson 43: 259
125. Macura S, Wüthrich K, Ernst RR (1982) J Magn Reson 46: 269; Rance M, Bodenhausen G, Wagner G, Wüthrich K, Ernst RR (1985) J Magn Reson 62: 497
126. Marion D, Zasloff M, Bax A (1988) FEBS Lett 227: 21
127. Kline AD, Wüthrich K (1985) J Mol Biol 183: 503

128. Billeter M, Kline AD, Braun W, Huber R, Wüthrich K (1989) J Mol Biol 206, 677
129. Boelens R, Scheek RM, van Boom JH, and Kaptein R (1987) J Mol Biol 193: 213
130. Anderson JE, Ptashne M, Harrison SC (1985) Nature 316: 596
131. LeMaster DM, Richards FM (1988) Biochemistry 27: 142
132. Bothner-By AA, Stephens RL, Lee J, Warren CD, Jeanloz RW (1984) J Amer Chem Soc 106: 811
133. Bax A, Davis DG (1985) J Magn Reson 63: 207
134. Marion D (1985) FEBS Lett 192: 99
135. Kessler H, Griesinger C, Kerssebaum R, Wagner K, Ernst RR (1987) J Amer Chem Soc 109: 607
136. Griesinger C, Ernst RR (1987) J Magn Reson 75: 261
137. Wagner G, Wüthrich K (1982) J Mol Biol 160: 343
138. Wagner G (1983) Quart Rev Biophys 16: 1
139. Wagner G, Wüthrich (1986) Meth Enzym 131: 307
140. Anil-Kumar, Wagner G, Ernst RR, Wüthrich K (1980) Biochem Biophys Res Commun 96: 1156
141. Wider G, Hosur RV, Wüthrich K (1983) J Magn Reson 52: 130
142. Zuiderweg ERP, Hallenga K, Olejniczak ET (1986) J Magn Reson 70: 336
143. Redfield AG, Kuns SD, Ralph EK (1975) J Magn Reson 19: 114; Plateau P, Guéron M (1982) J Amer Chem Soc 104: 7310; Hore PJ (1983) J Magn Reson 55: 283
144. Patel DJ, Kozlowski SA, Weiss M, Bhatt R (1987) Biochemistry 24: 936
145. Kochoyan M, Leroy J-L, Guéron M (1987) J Mol Biol 196: 599
146. Hare DR, Ribeiro NS, Wemmer DE, Reid BR (1985) Biochemistry 24: 4300
147. Hilbers CW, Heerschap A, Hassnoot CAG, Walters JALI (1983) J Biomolec Struct Dyn 1: 183
148. Sklenář V, Bax A (1987) J Magn Reson 74: 469
149. Otting G, Grütter R, Leupin W, Minganti C, Ganesh KN, Sproat BS, Gait MJ, Wüthrich K (1987) Eur J Biochem 166: 215
150. Sklenar V, Brooks BR, Zon G, Bax A (1987) FEBS Lett 216: 249
151. Thomas MA, Anil-Kumar (1985) J Magn Reson 61: 540
152. Bolton PH, Bodenhausen G (1982) Chem Phys Lett 89: 139
153. Widmer H, Wüthrich K (1986) J Magn Reson 70: 270
154. Rowan R III, Sykes BD (1975) J Amer Chem Soc 97: 1023
155. Brisson J-R, Carver JP (1983) Biochemistry 22: 1362
156. Keller J, Neuhaus D, Williamson MP (1987) J Magn Reson 73: 45
157. Kay LE, Scarsdale JN, Hare DR, Prestegard JH (1986) J Magn Reson 68: 515
158. Kay LE, Holak TA, Johnson BA, Armitage IM, Prestegard JH (1968) J Amer Chem Soc 108: 4242
159. Kojiro CL, Markley JL (1983) FEBS Lett 162: 52
160. Frey MH, Wagner G, Vasak M, Sørensen OW, Neuhaus D, Wörgötter E, Kägi JHR, Ernst RR, Wüthrich K (1985) J Amer Chem Soc 107: 6847
161. Paulsen H, Peters T, Sinnwell V, Heume M, Meyer B (1986) Carbohydrates Res 156: 87
162. Lerner L, Bax A (1987) Carbohydrates Res 166: 35
163. Bock K, Pedersen C, Pedersen H (1984) Adv Carbohydr Chem Bioch 42: 193
164. Bodenhausen G, Freeman R (1977) J Magn Reson 28: 471
165. Freeman R, Morris GA (1978) J Chem Soc Chem Commun 684
166. Bax A, Morris GA (1981) J Magn Reson 42: 501
167. Levitt MH, Freeman R, Frenkiel T (1983) Adv Magn Reson 11: 48
168. Bendall MR, Pegg DT, Doddrell DM (1983) J Magn Reson 52: 81
169. Bax A, Griffey RH, Hawkins BL (1983) J Magn Reson 55: 301
170. Garbow JR, Weitekamp DP, Pines A (1982) Chem Phys Lett 93: 504
171. Chan TM, Markley JL (1982) J Amer Chem Soc 104: 4010
172. Markley JL, Westler WM, Chan TM, Kojiro CL, Ulrich EL (1984) Fed Proc 43: 2648
173. Kessler H, Bermel W, Griesinger C (1985) J Amer Chem Soc 107: 1083
174. Kessler H, Griesinger C, Lautz J (1984) Angew Chem Int Ed Engl 23: 444
175. Kessler H, Müller A (1986) Liebigs Ann Chem 1687
176. Otter A, Kotovych G (1987) J Magn Reson 74: 293
177. Bax A, Summers MF (1986) J Amer Chem Soc 108: 2093
178. Bolton PH (1982) J Magn Reson 48: 336
179. Kessler H, Bernd M, Kogler H, Zarbock J, Sørensen OW, Bodenhausen G, Ernst RR (1983) J Amer Chem Soc 105: 6944
180. Lerner L, Bax A (1986) J Magn Reson 69: 375
181. Braun W (1987) Quat Rev Biophys 19: 115
182. Brünger AT, Clore GM, Gronenborn AM, Karplus M (1986) Proc Natl Acad Sci 83: 3801

183. Kaptein R, Zuiderweg ERP, Scheek RM, Boelens RM, van Gunsteren WF (1985) J Mol Biol 182: 179
184. LeMaster DM, Richards FM (1985) Biochemistry 24: 7263
185. Griffey RH, Redfield AG, Loomis RE, Dahlquist FW (1985) Biochemistry 24: 817
186. Weiss MA, Redfield AG, Griffey RH (1986) Proc Natl Acad Sci 83: 1325
187. Otting G, Senn H, Wager G, Wüthrich K (1986) J Magn Reson 70: 500
188. Bax A, Weiss MA (1986) J Magn Reson 71: 571
189. Fesik SW, Gampe RT, Rockway TW (1987) J Magn Reson 74: 366
190. Griffey RH, Redfield AG (1987) Quat Rev Biophys 19: 51
191. Cooley JW, Tukey JW (1965) Math Comput 19: 297
192. Ernst RR (1966) Adv Magn Reson 2: 1
193. Barkhuijsen H, de Berr R, Bovee WMMJ, van Ordmondt D (1985) J Magn Reson 61: 465
194. Delsuc MA, Ni F, Levy GC (1987) J Magn Reson 73: 548
195. Laue ED, Skilling J, Staunton J, Sibisi S, Brereton RG (1985) J Magn Reson 62: 437
196. Laue ED, Mayger MR, Skilling J, Staunton J (1986) J Magn Reson 68: 14
197. Griesinger C, Sørensen OW, Ernst RR (1987) J Amer Chem Soc 109: 7227
198. Dubs A, Wagner G, Wüthrich K (1979) Biochim Biophys Acta 577: 177
199. Lehming N, Sartorius J, Niemoller M, Genenger C, v. Wilcken-Bergmann B, Müller-Hill B (1987) EMBO J 6: 3145
200. Oschkinat H, Griesinger C, Kraulis PJ, Sørensen OW, Ernst RR, Gronenborn AM, Clore GM (1988) Nature (London) 332: 374
201. Vuister GW, Boelens R, Kaptein R (1988) J Magn Reson 80: 430
202. Marion D, Kay LE, Sparks SW, Torchia DA, Bax A (1989) J Amer Chem Soc 111: 1515; Kay LE, Marion D, Bax A (1989) J Magn Res 84: 72
203. Zuiderweg ERP, Fesik SW (1989) Biochemistry 28: 2387
204. Fesik SW, Zuiderweg ERP (1988) J Magn Reson 78: 588

Solid-State NMR in High and Very High Magnetic Fields

U. Haeberlen

Max-Planck-Institut für Medizinische Forschung, Arbeitsgruppe Molekülkristalle, Heidelberg, FRG

Table of Contents

This contribution grew out of a seminar held in Grenoble in September '86 on the prospects of the construction of a 1-GHz NMR spectrometer whose crucial component is, of course, the magnet which is planned to generate a field of 23.4 Tesla. My part in this seminar was, as it is in this volume, to consider solid-state NMR in fields currently classified as "high" as well as to explore the impact the availability of a 23.4-Tesla field would possibly have on this branch of NMR spectroscopy. Naturally I shall not restrict myself to proton NMR to which the frequency of 1 GHz applies but shall keep an eye on the NMR of other nuclei as well. What follows is not intended to be a review of solid-state NMR as such, or of solid-state NMR methodology. Such reviews with extensive lists of references have appeared in the literature in various places [1–4].

NMR Basic Principles and Progress, Vol. 25

1 Introduction

NMR in solids is characterized, unlike liquid-phase NMR, by an extraordinarily large diversity of experiments people wish to perform. The motivation for doing many of these experiments is to elucidate basic phenomena characteristic of solids or, e.g., very low temperatures. Somewhat randomly selected examples are the approach to thermal equilibrium of an initially nonequilibrium spin state via high-order multiple quantum coherences, spin diffusion in general, the role of Berry's phase under various circumstances, spin conversion in solid-state quantum systems such as solid methane and hydrogen, and the fascinating behavior of superfluid and solid ^{3}He. The various schemes designed to create non-thermal, often very large nuclear spin polarisation by optical and microwave irradiation also belong to this category.

Nevertheless the purpose of the vast majority of solid-state NMR experiments and measurements is to probe matter by one or the other of more or less standard NMR techniques. Thus, the electronic properties of semiconductors, normal metals as well as, recently, organic metals, classical superconductors as well as, recently, organic and ceramic (so-called high-T_c) superconductors are studied by relaxation (T_1) and Knight-shift measurements as is the diffusive and tunneling motion of protons/deuterons and small ions in metals and, often, in superionic conductors. A large field for NMR investigations concerns the structure and dynamics (e.g., phase transitions) of ionic crystals. In these systems it is often the interaction of electric field gradients with quadrupolar nuclei which is exploited. A special branch with facets of its own is represented by incommensurate systems and their elementary excitations, phasons, and amplitudons. In recent years people realize more and more that glassy systems pose altogether new problems and some investigators consider them even as being one of the few remaining basic challenges of solid-state physics. NMR is certainly one of the most powerful tools for the study of glassy systems.

Another class of systems that should be mentioned is that of molecular crystals of organic compounds. NMR is particularly suited to study localized motions in such crystals, i.e., the jumpwise and, at very low temperatures, tunneling reorientations of methyl groups ($—CH_3$, $—CD_3$) and of ammonium ions (NH_4^+, ND_4^+), the 180°-flips of H_2O (D_2O) molecules in hydrates, and the dynamics of hydrogen bonds which plays a crucial role in a vast diversity of solids such as concrete, the technically important ferroelectrics, crystals of carboxylic acids and their salts and, on the other end of the scale, DNA and proteins.

While all these different kinds of systems can be studied successfully by NMR, the equipment used is typically "homebuilt" and specifically tailored to the particular case. The question about the optimum strength of the applied field B_0 must be asked separately in each case and the strongest field available "on the market" is by no means automatically the preferred one. For some problems it may even turn out that zero-field NMR is the solution.

On the other hand, there are classes of materials and types of experiments which are studied by solid-state NMR on a more-or-less routine basis, and with standardized, commercial, highly perfectioned equipment. To these materials belong, in particular, plastics and polymers, zeolites, and inclusion compounds. The outstanding single one solid-state NMR technique on which, because of its wide applicability, a tremendous amount of technical efforts has been spent, is magic angle sample spinning (MASS), often combined with cross polarization and heterodecoupling. It is applied mainly to ^{13}C, ^{15}N, ^{29}Si, and ^{27}Al nuclei. Deuterium NMR in many forms and up to three dimensions has evolved in the last decade as one of the most powerful spectroscopic techniques for the study of polymers.

In what follows, I shall consider the impact which the construction of ever stronger magnets has had on these areas in the past, and what could foreseeably be gained if a 23.4-Tesla or even stronger magnetic field was available as it will be, we all hope, in the near future. This impact is different from that the advent of high-field superconducting solenoids has had on liquid-state and biological NMR. In these branches of NMR spectroscopy the progress is mainly connected with sensitivity and increase of dispersion. The latter aspect is due to the paramount importance of isotropic chemical shifts in liquid-state NMR.

In solid-state NMR, by contrast, the role of isotropic chemical shifts is much more modest but has been quickly increasing in the last few years. In solids we often exploit dipolar and/or quadrupolar interactions of the spins as well as the anisotropic part of the nuclear shielding to gain information about the sample under study. Taking very low fields apart, the NMR spectral features of dipolar interactions are independent of the strength of B_0 and in this respect we do not gain by using higher fields. For quadrupolar nuclei and their interactions with electric field gradients the situation is more complicated: depending on the nucleus considered and the particular case, the application of a higher field may simplify as well as complicate the spectra (see below). At high fields, chemical shifts may become comparable in size with dipolar and quadrupolar interactions which is, on the one hand, a blessing because of the potentially increased information content of the spectra but may be, on the other hand, a curse because of increased spectral complexity. I shall quote a specific example further below.

The importance of relaxation studies is much higher in solid than in liquid-state NMR. People have learned to appreciate more and more the value of the information contained in the frequency or field dependence of nuclear relaxation rates. The range of interest for the variation of the field strength encompasses usually several orders of magnitude. Of particular interest are often very low rather than very high fields. This led consequently to the design and construction of so-called field-cycling spectrometers. In this area an increase of the maximum field by a factor of two or so is hardly considered as a breakthrough.

The diversity of the experiments and the modest role of the chemical shifts in solid-state NMR are the main reasons why solid-state NMR spectroscopists have traditionally not been in the foremost front of those who pushed builders of magnets and spectrometers to provide ever higher fields. They have rather tended to be in the "second wave" and have profited later on from what had been developed for liquid-state and biological NMR. Even now when "500 MHz" spectrometers ($B_0 = 11.74$ Tesla) are routine in biological NMR, and 600 MHz systems are being introduced, there are only very few solid-state NMR labs which have a 11.74 Tesla magnet available for their work. If a decision needs to be made between a high field and a large freely accessible volume (bore diameter in the typical case of a superconducting solenoid), solid-state NMR spectroscopists usually give higher priority to the latter aspect because in addition to the NMR probehead proper, they often want to put additional gadgets inside the magnet such as a helium cryostat, a high pressure unit, a drawing device to study the elongation of polymers, etc. This priority list is exemplified by a recent decision taken at the Max-Planck-Institut for Polymer Research in Mainz: A very wide-bore 7-Tesla magnet was given preference to a standard 11.74-Tesla solenoid.

2 Sensitivity Considerations for High and Very High Field Solid-State NMR

One of the prime motivations for constructing a very high field magnet and the corresponding NMR spectrometer is *sensitivity*. In this section I shall discuss the special facts of NMR sensitivity in solids.

In liquid-phase NMR the inverse of the bandwidth of the NMR probe, or its quality factor Q, is usually at its technical upper limit. The experiment itself would often tolerate a still higher Q. This is not so in solid-state NMR. The width of the spectrum to be recorded, or the required impulse response of the probe call for a quality factor Q at some lower value than the technically feasible maximum. Examples are deuterium and high-resolution multiple-pulse proton NMR. The typical width of a deuterium spectrum is 300 kHz. For the recording of spectra free of distortions, especially of powder and 2D-spectra, a bandwidth $\Delta\nu$ of the probe of the order of 2 MHz is required. This corresponds to $Q = 25$ at current deuterium frequencies of about 50 MHz. In proton multiple-pulse NMR we wish to record the FID no later than 2 μs after the beginning of a $\pi/2$ pulse. This calls for a bandwidth $\Delta\nu$ of the probe in excess of 10 MHz.

Because of this extra design constraint, there are modifications of the field dependence of the solid-state NMR sensitivity in comparison to that of liquid-state NMR. Figure 1 shows schematically the electrical circuit of an NMR probe.

The precessing nuclear magnetization induces a voltage U_{ind} in the NMR coil whose inductance is L. The coil is supplemented by C and R_i to make a series

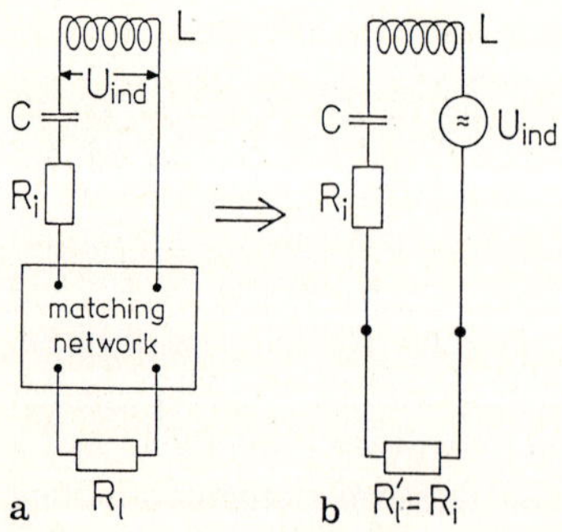

Fig. 1a and b. NMR probe circuit. **a.** real; **b.** simplified equivalent version

resonance circuit with $Q = \omega_o L/R_i$. The preamplifier input resistance plays the role of the load resistance R_l. For commercial high frequency, low-noise preamplifiers, the input resistance is usually $R_{in} = 50\,\Omega$ if the amplifier is to be used in a $Z_o = 50\,\Omega$ system. Such amplifiers are internally matched to the optimum noise resistance of the first active element.

As R_i is usually much smaller than Z_0, a matching network is used to transform R_i "up" to Z_0. Ideally this network is lossless and in practice it is nearly so. The power N_l delivered to R_l in the actual case is therefore identical to that delivered to $R_l' = R_i$ in the simplified version of the circuit shown in Fig. 1b, in which we have also separately illustrated the source of the signal. In this form, the internal resistance of the signal source is zero. The "loaded" quality factor Q_l of the circuit (b) is $Q_l = \omega_o L/(2R_i)$. It is Q_l which determines the bandwidth. Under matching conditions the power delivered to R_l' (Fig. 1b), or to R_l (Fig. 1a), is:

$$N_l' = N_l = U_{ind}^2/4R_i = U_{ind}^2 \frac{Q_l}{2\omega_o L}$$

The signal voltage S, i.e., the voltage at the input of the preamplifier is:

$$S = \sqrt{N_l Z_o} = U_{ind} \sqrt{\frac{Q_l Z_o}{2\omega_o L}} \tag{1}$$

According to the reciprocity theorem (Hoult and Richards, 1976) the induced voltage immediately after a $\pi/2$-pulse of a NMR sample of volume V_s and spin density N_s which had been in thermal equilibrium ($M_z = M_o$) prior to the pulse is:

$$U_{ind} = V_s M_o \omega_o \frac{B_{trans}}{I} \tag{1a}$$

where B_{trans} is the magnetic field component perpendicular to B_o which a current I through the probe coil generates at the site of the sample. We assume that B_{trans} is homogeneous over the sample volume. As

$$M_o = \frac{N_s \gamma^2 \hbar^2 I(I+1) B_o}{3kT} = N_s \frac{(\hbar\omega_o)^2}{4kTB_o} \quad \text{for } I = 1/2$$

we get for the signal voltage:

$$S = N_s V_s \frac{(\hbar\omega_o)^2}{4kT} \frac{\omega_o}{B_o} \frac{B_{trans}}{I} \sqrt{\frac{Q_l Z_o}{2\omega_o L}} \tag{2}$$

S seems to depend on the inductance L of the coil. Actually it does not, and in order to display this we specialize to the case of a (long) cylindrical coil for which:

$$B_{trans}/I = \mu_o n/l \quad \text{and} \quad L = \mu_o (n/l)^2 V_c$$

where n, l, and V_c are, respectively, the number of turns, the length, and the volume of the coil. Inserting these expressions into Eq. (2) we get:

$$S = N_s V_s \frac{(\hbar\omega_o)^2}{4kT} \frac{\omega_o}{B_o} \sqrt{\frac{Q_l Z_o \mu_o}{2\omega_o V_c}} \tag{3}$$

For other types of coils a "geometry factor" which generally is smaller than one, has to be added as a factor in the r.h.s. of Eq. (3). According to this expression, the signal voltage S seems to increase with the power 3/2 of ω_o. However, if the experiment calls for a definite, frequency-independent bandwidth $\Delta\omega = 2\pi\Delta\nu = \omega_o/Q$, it is appropriate to rewrite S in the following form:

$$S = N_s V_s \frac{(\hbar\omega_o)^2}{4kT} \frac{\omega_o}{B_o} \sqrt{\frac{\mu_o Z_o}{2\Delta\omega V_c}} \tag{3a}$$

which shows that for this case S grows with the second power of ω_o. This means that the sensitivity of solid-state NMR benefits fully from the "Curie" factor (M_o) as well as from the "Faraday" factor (ω_o) on increasing the strength of the static applied field B_o.

Another useful expression for the peak signal voltage S is obtained by combining Eqs. (1) and (1a) with the obvious equations:

$$2B_1 = (B_{trans}/I) \cdot I_{tr}$$

$$N_{tr} = I_{tr}^2 \cdot \frac{R_i}{2} = I_{tr}^2 \cdot \frac{\omega_o L}{4Q_l} \quad \text{and}$$

$$\gamma B_1 t_{\pi/2} = \pi/2$$

where N_{tr} = power delivered by the transmitter to the probe,
I_{tr} = peak current flowing through R_i and L during the pulse,
B_1 = rotating rf-field strength during the pulse, and
$t_{\pi/2}$ = duration of a 90° pulse obtained with transmitter power N_{tr}.

We get:

$$S = V_s M_o B_o \frac{\pi}{2\sqrt{2}} \sqrt{\frac{Z_o}{N_{tr}}} \frac{1}{t_{\pi/2}}$$

which reduces for $I = 1/2$ to:

$$S = V_s N_s \frac{(\hbar\omega_o)^2}{4kT} \frac{\pi}{2\sqrt{2}} \sqrt{\frac{Z_o}{N_{tr}}} \frac{1}{t_{\pi/2}}$$

All quantities referring to the coil geometry such as V_c or L have canceled out! That this is possible is a consequence of the reciprocity theorem. What makes these expressions so pleasing is the fact that S is given in terms of fundamental constants ($k, \hbar$) and quantities which are either well known for a given spectrometer (Z_o, ω_o) or are easily measured ($N_{tr}, t_{\pi/2}, T$) and finally the number $N = V_s N_s$ of resonating spins in the sample.

The noise voltage N appearing at the input of the preamplifier is equal to:

$$N = \sqrt{4kTZ_o\Delta\nu} = \sqrt{\frac{2}{\pi} kTZ_o\Delta\omega} \tag{4}$$

Because of the matching network, it is Z_o which appears in Eq. (4) at the place of the physical noise-generating resistance. The signal-to-noise ratio:

$$S/N = N_s V_s \frac{(\hbar\omega_o)^2}{(4kT)^{3/2}} \frac{\omega_o}{B_o} \frac{1}{\Delta\omega} \sqrt{\frac{\mu_o \pi}{V_c}} \tag{5}$$

is independent of Z_o, as it must. Equation (5) also shows that S/N varies (in a well-designed apparatus!) as the inverse 3/2 power of the temperature T, a fact that can be exploited much more readily with solid than, in particular, biological samples.

The last factor influencing the sensitivity of a solid-state NMR experiment which we have to consider is the spin-lattice relaxation time T_1. The range of T_1's encountered in solids is enormously large, much larger than in liquids. For quadrupolar nuclei in mobile surroundings T_1 can well be shorter than 1 ms. Further below we shall specifically consider the case of deuterons. In rigid crystals, on the other hand, T_1 is often as long as several hours. This is true of spin-1/2 and also of quadrupolar nuclei. In view of these huge differences in T_1 the increase of the spin lattice relaxation time which we have to expect if B_o is increased from, say, 8 to 23.4 Tesla is hardly worth mentioning. In most cases this increase can easily be counteracted by changing the sample temperature, either up or down.

This means that from a twofold increase of the strength of the applied field, solid-state NMR benefits rather generally by a fourfold increase of the sensitivity. There can be no doubt that an increase of the accessible field to 23.4 Tesla would considerably (further) enlarge the number of chemical and physical problems that can be solved by solid-state NMR. In single crystal work, increased sensitivity means that work can be performed with smaller crystals. In high-resolution solid-state NMR, increased sensitivity renders accessible more complex systems. The solid-state NMR of weak, rare nuclei becomes practical and *surface science* in particular would greatly benefit from any substantial progress in solid-state NMR sensitivity.

3 Solid-State Proton NMR

This field may be classified into two categories: Wide-line and high-resolution solid-state proton NMR which calls for applying line-narrowing multiple pulse sequences. High resolution methods allow to observe separate resonances from magnetically inequivalent sites. When the multiple pulse sequences are supplemented by magic angle sample spinning—which makes sense in particular in work on powder samples—separate resonances are observed from crystallographically inequivalent sites. In wide-line proton NMR, individual resonances from different sites are not resolved. In regard to relaxation, spin-diffusion causes, as a rule, all protons to relax with the same relaxation time towards thermal equilibrium, irrespective of whether or not they are in equivalent sites. Relaxation time measurements may therefore as well be encompassed by wide-line NMR.

3.1 Consideration of the Strength of B_0 in Wide-Line Proton NMR

There is, of course, the usual question about sensitivity. It is, however, of rather little importance in wide-line proton NMR since, on the one hand, protons are anyway the most sensitive NMR nuclei, and on the other hand, all protons of the sample contribute to the same resonance. Next we must consider the field dependence of the shape of the resonance. This shape is dominated by the dipolar interactions among the proton spins. The information contained in these interactions about molecular and crystal structure and about molecular dynamics are of central interest in wide-line proton NMR. Line-splittings specific of pair interactions can be resolved only in rare cases involving single crystals of simple molecules. Two examples are malonic and paratoluic acid [5, 6]. Usually one must be content with a moment analysis of the spectra which actually is restricted to the 2nd and, perhaps 4th moment. For a given molecular and crystal structure the contribution of the direct dipolar spin-spin interactions to these moments can be calculated. If the dipolar contributions dominate the total moments, the calculated ones can be compared directly with experimental results. This situation prevails at medium fields ($B_0 \lesssim 3$ Tesla) where chemical shifts can safely be neglected in comparison to dipolar line broadening. At high fields, however, chemical shifts give a sizeable contribution to the 2nd and fourth moments, etc. and cause the first and other odd moments of the NMR to become nonzero.

A theoretical interpretation of the experimental moments in terms of molecular and crystal structure then becomes close to impossible. We have encountered such a situation in the course of a ^{19}F-NMR study of SnF_2 using a field of 2.14 Tesla [7]. As ^{19}F-shifts and shift anisotropies are roughly ten times as large as those of hydrogens, the disturbing effect of chemical shifts/shift anisotropies on dipolar structured wide-line spectra is felt already at correspondingly smaller fields in ^{19}F NMR. In the prospected field of 23.4 Tesla these effects will be very serious also in wide-line proton NMR.

Our conclusion then is that "high" and "very high" fields are less suitable for proton wide-line NMR than are fields around, say, 2 Tesla. Also it should be clearly outspoken that while it is very difficult to assess the value of any work in basic research in terms of cost, there seems to exist agreement that the scientific yield of very high field wide-line proton NMR does not justify the very high costs involved. This is certainly true of standard wide-line proton NMR. However, it does not exclude the possible existence of important scientific problems whose solutions may be obtained by a very high field wide-line proton-NMR experiment and where costs should stay out of consideration.

3.2 High-Resolution Solid-State Proton NMR

This line of research dates back to 1967 when J.S. Waugh, L. Huber, and the present author showed that certain cyclic multiple pulse sequences have the effect of suppressing effectively the line-broadening dipolar homonuclear spin-spin interactions while chemical shifts are, at the same time, retained. This opened up the possibility to measure chemical shift tensors of protons, ^{19}F, and other nuclei. In view of the tremendeous importance of isotropic shifts, expectations were high when the full tensors became accessible to measurement.

The first successful experiment was carried out on the ^{19}F nuclei in CaF_2 in a field corresponding to a ^{19}F-Larmor frequency of 54 MHz. Several reasons—the least of them sensitivity—led to the application of ever higher fields in this branch of NMR spectroscopy. The two most important ones are:

i) The unwanted effect, i.e., the dipolar line broadening which must be suppressed, is independent of the strength of the applied field. The sought-after effects, i.e., the chemical shifts and shift anisotropies increase, on the other hand, in proportion to B_o. Any increase in B_o renders, therefore, the ratio of the sought-after/unwanted effects more favorable.
ii) The line-narrowing effect of the multiple pulse sequences is based on averaging in spin space. Therefore the sequences should work the better, the faster they are run and they actually do, albeit not to the extent theoretically expected. Running a multiple pulse sequence at a given speed requires a certain bandwidth of the rf-transmitter and of the NMR probe. We give an example: it is desirable to run, e.g., the so-called WAHUHA sequence consisting of four $\pi/2$-pulses spaced at intervals τ and 2τ with a total cycle time $t_c = 6\tau = 12\,\mu s$. This leaves a "window" of $4\,\mu s$ including one pulse width to detect the NMR. To achieve this, a bandwidth $\Delta\nu$ of roughly 10 MHz is required. The quality factor Q of the probe is therefore limited by bandwidth constraints and not by technical limitations. At higher fields and higher Larmor frequencies the same bandwidth is obtained with a higher quality factor Q. Higher Q's are preferable with regard to transmitter power and sensitivity and higher fields should, therefore, be advantageous.

In view of these considerations it was but natural that the strength of the applied field was increased for this type of experiments in the course of time. Our group did the steps 54 MHz $\Rightarrow$ 90 MHz $\Rightarrow$ 270 MHz. To the best of my knowledge commercial spectrometers with line-narrowing multiple pulse capabilities are now offered with Larmor frequencies up to 400 MHz.

Did the results obtained at higher fields live up to the expectations and what are the prospects if ν_o is pushed to 1 GHz?

When we did the step from 90 MHz to 270 MHz we were disappointed to see that under nominally the same conditions, i.e., equal cycle times and equal pulse widths, the linewidths in Hz were *larger* at 270 MHz than at 90 MHz. True, they were *narrower* in terms of ppm, i.e., a net gain remained. Nevertheless it was smaller than hoped for. I should stress that other groups obtained similar results.

One reason for this disappointing finding is the existence of at least one line-broadening mechanism which is proportional to B_o: Warping of the applied field by bulk susceptibility inhomogeneity. In liquid-state NMR this problem is overcome by using samples in the form of long cylinders of which only the central part is made NMR-active. This approach is not feasible in high resolution solid-state proton NMR since the necessity for applying a homogeneous B_1 field requires the entire sample crystal to be well inside the rf-coil. Samples in the form of long cylinders are therefore not usable. What remains to prevent bulk susceptibility inhomogeneity from warping the field inside the sample is the use of *spherical* samples. While it is easy to grind a sphere out of a hard crystal with cubic symmetry such as CaF_2, it is very difficult to shape spheres out of triclinic and monoclinic crystals. Remember that the vast majority of compounds in nature crystallize in the monoclinic system. On several occasions we made the disappointing experience that the tendency of a crystal to split along one of its easy cleavage planes rises sharply as the shape of the specimen approaches that of a sphere! Even when we succeed to get a sphere of the sample crystal, there remains the problem of mounting it in the NMR sample tube in a given orientation in a way that does not reintroduce warping of the field. Several authors have suggested to immerse the crystal in a fluid with a susceptibility χ_{fluid} matching that of the sample crystal. In general this will not work since the crystal's susceptibility χ_{crystal} will be anisotropic, whereas that of the fluid is not. The matching is therefore lost when the crystal is rotated in the fluid, as it must in a full experiment. A measure for the size of the field warping effect due to bulk susceptibility inhomogeneity is the size of χ itself. For many organic and ionic crystals $\chi \approx 0.7$ ppm. This means, whenever a resolution of better than 1 ppm is desired, the field warping due to bulk susceptibility inhomogeneity must be seriously considered.

The prime technical difficulty which I anticipate for multiple pulse NMR at 1 GHz is the requirement of a strong, homogeneous rf-field over the sample volume. In a sphere of, say, 4 mm diameter, $\sqrt{\langle \delta B_1 \rangle^2_{av}}/B_1$ should be less than 0.02. To construct a small-size rf-antenna at 1 GHz meeting this requirement appears to be a real challenge, to say the least.

Finally we should again raise the question of whether or not it may be worthwhile in terms of scientific yield to perform multiple-pulse line-narrowing measurements at 1 GHz. Certainly the answer cannot be an enthusiastic "yes". If it was "yes", high-resolution solid-state NMR spectroscopists would have made more efforts, and more successful ones than they actually did, to make available for their experiments 500-MHz or 600-MHz systems. One reason for this may be that the measurement of a chemical shift tensor is a rather tedious undertaking even if the experiment works close to perfection. In any event, it requires growing of single crystals, orienting them by optical or X-ray means and recording many spectra (typically ≈ 50). This is too much effort to make chemical shift tensors valuable for analytical purposes. Inspite of this judgement I hope and assume that in some decades when the technical progress has gone beyond the 1 GHz limit for solution high-resolution NMR, multiple-pulse line-narrowing experiments will be done quite routinely at 1 GHz as well.

4 Solid-State Deuteron NMR

Several research groups around the world nowadays apply solid-state deuteron NMR on a more-or-less routine basis to the study of structure and dynamics in biological systems (e.g., membranes), polymers, inclusion compounds, and in ionic and molecular crystals. Cryomagnets with fields $B_o = 4.7 \ldots 9.4$ Tesla are usually applied. These correspond to deuteron Larmor frequencies $\nu_o = 30.7 \ldots 61.4$ MHz. This is to say that the highest fields currently available are typically not used for solid-state deuteron NMR. The extra costs for the higher fields is the only reason for not using them. Ignoring costs, an increase of the applied field to the prospective value of 23.4 Tesla appears to be particularly attractive for solid-state deuteron NMR for several reasons. First, the deuteron Larmor frequency in such a field is 153.5 MHz. It falls in a range where the required rf-technology is well developed and readily available. Also in this frequency range there exists a wealth of experience with regard to construction of NMR probes. Second, the field homogeneity and stability requirements for this type of work are less stringent than for high-resolution NMR proper: The width of the narrowest deuteron NMR lines even in single crystals where the deuterons are magnetically fairly well isolated, is hardly less than 1 kHz. Therefore, a basic linewidth of, say, 100 Hz to be obtained with a liquid test sample is perfectly adequate with regard to the magnet's homogeneity and similar considerations apply to the stability of the field.

4.1 Deuteron NMR Sensitivity

For an assessment of the NMR sensitivity, a consideration of the spin-lattice relaxation times T_1 is important. The T_1's of deuterons in solids vary considerably although at least in diamagnetic solids they are dominated invariably by the interaction of the nuclear quadrupole moment with the fluctuating part of the electric field gradient (EFG) at the site of the nucleus. As the electric field gradient at the site of a deuteron is almost exclusively a property of the chemical bond, the EFG will fluctuate when the bond reorients. A rough guess of the *shortest* relaxation time $T_{1\,\min}$ is obtained from the expression of $1/T_1$ of a deuteron whose bond carries out random jumps between two energetically equivalent sites. In this process the EFG jumps from $\mathbf{V}_i$ (i for initial) to $\mathbf{V}_f$ (f for final) where $\mathbf{V}_f$ is obtained by a rigid rotation of $\mathbf{V}_i$. The calculation yields [8]:

$$1/T_1 = \frac{4}{9}\pi^2 \left\{ [(Q^{fl}_{xx} - Q^{fl}_{yy})^2 + 4(Q^{fl}_{xy})^2] \frac{\tau_c}{1 + 4\omega_o^2 \tau_c^2} + [(Q^{fl}_{xz})^2 + (Q^{fl}_{yz})^2] \frac{\tau_c}{1 + \omega_o^2 \tau_c^2} \right\}$$

where

$$\mathbf{Q}^{fl} = \frac{3}{4}\frac{eQ}{h}(\mathbf{V}_i - \mathbf{V}_f).$$

τ_c is the correlation time of the jumps. Naturally, the relaxation rate is dependent upon the orientation of the applied field B_o relative to $\mathbf{Q}^{fl}$. The z-axis of the reference frame is parallel to B_o. If $\mathbf{V}_i$ is axially symmetric, as it approximately is for deuterons, a situation can be found by rotating the crystal where $Q^{fl}_{xy} = Q^{fl}_{yx}$ is the only nonzero tensor-component of $\mathbf{Q}^{fl}$. Remember that $\mathbf{Q}^{fl}$ is not a general second rank tensor, it is the difference of two traceless, axially symmetric tensors which differ only in their orientations, otherwise it would not be possible to find such an orientation! For this special orientation the size of Q_{xy}^{fl} is given by $\frac{3}{2}\Delta\nu_{max}$ where $\Delta\nu_{max}$ is the maximum splitting of the deuteron resonance in case it senses a *static* field gradient tensor $\mathbf{V}_i$. For a carbon-bonded deuteron, a typical value for $\Delta\nu_{max}$ is 270 kHz, corresponding to a quadrupole coupling constant of 180 kHz. For the optimum correlation time $\tau_{c\,opt} = 1/2\omega_o$ we get for $\omega_o = 2\pi.50$ MHz: $T_1 \approx 0.5$ ms.

In several compounds including the $-ND_3$ deuterons of triglycine sulfate [9] and the $-OD$ deuterons of deuterated ethanol in Dianin's inclusion compound, we have actually observed deuteron relaxation times shorter than 4 ms, which demonstrates that our rough estimate of the minimum deuteron relaxation time in solids is realistic. Both these numbers (4 ms and 0.5 ms) are so short that they will pose no limitations in any accumulation experiment. The computer whose job it is to do the addition/substraction of the data sets will be the bottleneck. Our numbers imply, on the other hand, that spin-lattice relaxation will hardly ever be a cause for limited resolution in solid-state deuteron NMR.

Deuterons in essentially immobile systems relax despairingly slowly. An example is crystalline deuterated anthracene whose T_1 at room temperature is $1\frac{1}{2}$ hours and it gets even longer when the temperature is lowered!

In view of this huge range of spin-lattice relaxation times it hardly matters whether or not the field is raised from the currently preferred one of 7 Tesla to the prospected one of 23.5 Tesla. There will be systems with very short and, also, very long relaxation times in both fields. As, in addition, there seem to be no physical reasons why the deuteron linewidths should increase with an increase of the strength of the applied field, we may expect that the sensitivity of solid-state deuteron NMR will benefit fully, i.e., in proportion to B_o^2, from any effort to make available very high magnetic fields.

The natural abundance of ^2H is $1.5\,10^{-2}\,\%$. This is so low that practically all the current deuteron NMR work on solid materials is done on isotopically enriched samples. This state of affairs will not be altered with the prospective introduction of very high field magnets into NMR. The very need for isotopic enrichment in solid-state deuteron NMR should, however, not be considered solely as a handicap. It has as well a positive facet since deuteration means almost automatically *site-specific* deuteration. This specificity often results in a dramatic simplification which is essential for extracting the desired information about molecular structure and dynamics from the spectra. This is true regardless of whether we have in mind spectra from single crystals or from powders, or one-dimensional or two-dimensional spectra.

4.2 Quadrupolar, Dipolar, and Chemical Shift Interactions of Deuterons

The main features of deuteron NMR spectra of solid samples arise from the nuclear quadrupole moment—electric field gradient (EFG) interactions. The purpose of most of the deuteron NMR measurements is the study of these interactions because they tell us so much about molecular structure and dynamics. It is convenient to define a quadrupole coupling tensor $\mathbf{Q} = \frac{3}{2} \cdot (eQ/h)\mathbf{V}$ where e, Q, h, and $\mathbf{V}$ are, respectively, the elementary charge, nuclear quadrupole moment, Planck's constant, and the EFG tensor. This coupling causes a splitting of the deuteron NMR into a doublet. If the experiment is done in a high field, the splitting $\Delta\nu_Q$ is symmetric with respect to the pure Larmor frequency and is given simply by $\Delta\nu_Q = Q_{ZZ}$. Below we will state what a "high" field is in this sense. The index Z refers again to the direction of the applied field. By recording the orientation dependence of $\Delta\nu_Q$ (or of Q_{ZZ}) it is possible to determine $\mathbf{Q}$ including its orientation. The outstanding importance of the quadrupole coupling of deuterons arises from the fact that the quadrupole coupling tensor $\mathbf{Q}$ of a deuteron is an almost axially symmetric second rank tensor whose pseudo-axial symmetry axis coincides closely with the deuteron's bond direction. The study of deuteron quadrupole coupling tensors thus allows to measure bond directions of hydrogens and to monitor the reorientations of these bonds. The "size of $\mathbf{Q}$", i.e., its largest component, varies but very little for carbon-bonded deuterons from one compound to another. A typical value is 270 kHz. Any field B_o with $\nu_o = \gamma_D B_o/2\pi \gg 270$ kHz may be considered as "high", and in such "high" fields the spectrum is independent of B_o. The essential features of powder spectra are singularities and sharp edges. These allow the determination of the principal components of quadrupole coupling tensors. In case of presence of molecular dynamics they provide exceptionally direct information about kind and speed of molecular motions. Ultra-slow molecular reorientations occurring on the time-scale of tens of milliseconds to seconds are very successfully monitored by 2D-exchange spectroscopy [4].

The omnipresent dipolar interactions among the deuteron spins themselves, and between the deuteron and other spins, cause a broadening of the deuteron resonances in spectra of single crystals, and a rounding of the singularities and sharp edges of the powder spectra. As mentioned above, the dipolar line broadening $\delta\nu$ is rarely less than 1 kHz and as a rule it is about 2 kHz. It is thus two orders of magnitude smaller than the (maximum) quadrupole splittings. It is this relation between quadrupole splittings and dipolar linewidths that allow us to measure orientations of Q-tensors, and thus of bond directions, with a precision of better than 0.3°. It is worth pointing out that dipolar interactions do not affect the symmetry of deuteron spectra.

Chemical shifts lead to shifts of the centers of gravity of the quadrupolar split doublets. Chemical shift *differences*, regardless whether arising from isotropic or anisotropic shielding, lead to asymmetries in the solid-state deuteron spectra. The chemical shifts including shift anisotropies of carbon-bonded hydrogens (protons or deuterons) cover a range of roughly 10 ppm. The shift anisotropies $\Delta\sigma$ of hydrogens in hydrogen bonds are typically 20...30 ppm, a few cases with larger anisotropies are known, to my knowledge the "record holder" is the $(FHF)^-$ anion with $\Delta\sigma = 44$ ppm.

At a Larmor frequency of 50 MHz (typical current value), 10 ppm correspond to 500 Hz and this should be compared with the deuteron NMR linewidth $\delta\nu$ arising from the dipolar interactions of the deuteron spins. As mentioned above, it is rarely smaller than 1 kHz and, as a rule of thumb, it is 2 kHz. This means that with current magnet technology, chemical shifts are smaller than the linewidths. However, the centers of gravity of resonance lines can be determined, depending on the signal/noise ratio S/N, with much greater precision than just $\delta\nu$. A realistic estimate of the error ε with which the position of a resonance line in a spectrum can be determined is $\varepsilon = \pm 0.35 \cdot \delta\nu/(S/N)$ [10]. It is assumed that the line in question does not partially or fully overlap with other lines in the spectrum. In proton NMR the lines of differently chemically shifted nuclei overlap strongly, but in deuteron NMR the lines arising from nonequivalent nuclei are drawn apart by the strong quadrupolar interactions and their positions can be located within $\pm\varepsilon$. In sufficiently simple cases, deuteron chemical shift tensors can, therefore, be extracted from a set of deuteron spectra of single crystals. In a tour de force, the deuteron chemical shift tensors of glycine have been obtained in this way [10]. Clearly, a rise of the Larmor frequency to 153.5 MHz would drastically change the situation: chemical shift differences, and variations of chemical shifts with the orientation of the sample crystal would become about equal to the linewidths and, hence, easily measurable. This would make the technique a widely applicable one for the study of deuteron, or, more general, hydrogen chemical shift tensors. We point out that the otherwise difficult problem of assigning chemical shift tensors to hydrogen sites becomes a simple one in this technique.

For the set of lines in a set of spectra recorded for different crystal orientations for which a chemical shift tensor is determined, a quadrupole coupling tensor may be determined also almost automatically. This latter one and thus the chemical shift tensor as well, can usually be assigned easily and unequivocally to a hydrogen site on the basis of the well-established near-parallelism of the pseudo-axial symmetry axis of the quadrupole coupling tensor and the deuteron's valence bond direction.

4.3 Pending Problems of Very High Field Deuteron NMR

All aspects of very high field deuteron NMR in solids we have considered so far are definitely promising. There are, however, also some potential problems we have to mention. The first arises in measurements of T_1. The desirable initial non-equilibrium condition $M(0) = -M_0$ cannot be realized easily because the condition $\omega_1 \gg \Delta\omega_Q$ where $\omega_1 = \gamma_{deuteron} B_1$ and $\Delta\omega_Q = 2\pi\Delta\nu_Q$ cannot be met with current spectrometers and NMR probes. The limitation in increasing B_1 is eventually electrical breakdown in the tuning capacitor of the probe. Hence we prefer the initial condition $M(0) = 0$. Because of, again, insufficient rf field strength B_1, a single 90°-pulse does not generate perfectly this initial condition, starting from thermal equilibrium. A saturation pulse sequence must, hence, be used. This will result in $M(0) = 0$, but will, at the same time, generate quadrupolar order in the spin system unless the frequency of the saturating pulses matches exactly the nuclear Larmor frequency. The presence of quadrupolar order after the saturating sequence becomes evident in the spectrum calculated by FT of the FID after the "read" pulse: The two components of the doublet which should be symmetric, i.e., purely "absorptive", are asymmetric, in extreme cases even antisymmetric and this is not the result of improper phasing of the spectrum. Proper phasing can be adjusted easily by firing the "read" pulse under thermal equilibrium conditions, i.e., with negligibly small quadrupolar order in the system, and setting the phase corrections such that all lines in the spectrum appear in absorption. We encountered this type of problem during our investigation of the orientation dependence of the spin-lattice relaxation rate of the $-ND_3$-deuterons in single crystals of triglycine sulfate [9]. We solved it partly by using composite $\pi/2$-pulses. Now, when the spectrum consists of several chemically shifted doublets, as it usually does, it is not possible to meet the on-resonance condition simultaneously for all doublets. Therefore, when accurate values of T_1 are of interest, the measurements must be done separately for each doublet in the spectrum setting $\omega_{rf} = \omega_{Larmor}$ for each doublet in turn. This is obviously inconvenient and time-consuming and it is clear that this problem aggravates at higher fields and Larmor frequencies.

A similar problem arises in all measurements where quadrupolar echos are generated and Fourier transformed to get a spectrum. This is the universally applied technique in the study of powder samples. The echos are generated with a 90°_φ-τ-$90^\circ_{\varphi\pm 90^\circ}$-$\tau$-echo sequence where the index of the 90°-pulses indicates the rf-phase. It is important to stress that the sequence must be applied on-resonance. The same sequence applied off-resonance with $\Delta\omega_{off}\tau = \pi/2$ is essentially equivalent to an on-resonance sequence with both pulses having the same (or anti-) rf-phase. Such a sequence does not generate a quadrupolar echo.

Now, again, in a sample with magnetically inequivalent deuterons it is impossible to have $\Delta\omega_{off} = 0$ for all deuterons. In powder work the chemical shift anisotropy will lead to the same situation even when site-specifically deuterated samples are used. We point out that it is not the ratio $\Delta\omega_{off}/\delta\omega$ (with $\delta\omega = 2\pi \times$ linewidth) which matters, but the size of $\Delta\omega_{off}\tau$. To get an idea of the seriousness of the problem, assume $\Delta\omega_{off} = \Delta\sigma\omega_o$ where $\Delta\sigma$ is the chemical shift anisotropy of the deuterons under consideration. Setting the radio frequency at one edge of the chemical shift powder pattern the other edge will be out-of-resonance by just the amount $\Delta\sigma\omega_o$. With $\Delta\sigma = 20$ ppm (typical for hydrogen bonds) and $\omega_o = 2\pi.50$ MHz $\Delta\sigma\omega_o$ will be 2 kHz. The choice of $\tau = 250\,\mu s$ gives $\Delta\sigma\omega_o = \pi/2$, i.e., the situation where no echo is produced. It is evident that powder spectra derived from quadrupolar echos generated under such conditions will be heavily distorted. In current powder work aliphatic or aromatic deuterons, rather than deuterons in hydrogen bonds, are usually studied. Their shift anisotropies are of the order of 5...8 ppm and not 20 ppm. Moreover, a typical value for τ will be rather 25 μs than 250 μs. Under these conditions, the spectral distortions we are discussing can be and are ignored in most of the practical applications.

It is clear, however, that a more than threefold rise of the deuteron Larmor frequency from 50 MHz to 153 MHz will bring the seriousness of this problem for aliphatic and aromatic deuterons in the same range as it is currently for deuterons in hydrogen bonds which, incidentally, are seldom investigated by deuteron NMR in powder samples.

Nevertheless, I think it is appropriate to close this section by expressing the conviction that solid-state deuteron NMR spectroscopists will be clever and imaginative enough to overcome these pending problems of very high field deuteron NMR in due time. In the long run deuteron NMR in solids will profit a great deal from the development of very high field magnets.

5 High-Resolution Solid-State ^{13}C NMR

This is the technique which is now considered in the NMR community, especially by chemists, as *the* high resolution solid-state NMR technique per se. This is so because the spectra obtained are comparable to the more familiar liquid-state NMR spectra, i.e., they are dominated by the isotropic chemical shifts. Often they can be interpreted by mere inspection. They are usually recorded from powder samples which circumvents all the tedious problems involved in handling single crystals. This includes the advantage that the desired information is contained in a single spectrum rather than in a set of many spectra which usually must be analysed by computer. For brevity I am referring in this section to ^{13}C-NMR, but the discussion is equally valid for high-resolution solid-state NMR of other rare spin-1/2 nuclei, the most prominent of them being ^{29}Si.

I presume that the reader is familiar with the technique itself which includes cross-polarization, decoupling, and rapid magic angle sample spinning (MASS). Let us consider here the field-dependent aspects of the experiment. Currently traditional 60, 90, and 100 MHz spectrometers are being used, while cryomagnets are applied in the frequency range of 180–400 MHz.

5.1 Decoupling

The interaction which must be suppressed is the direct dipole–dipole interaction between the carbon and proton spins. The role of the carbon spins can also be assumed by ^{29}Si and other rare nuclei, while ^{19}F spins, e.g., can be substituted for the protons. Setting very low fields aside, the interaction in question is the I_zS_z or secular part of the full dipole–dipole Hamiltonian. It is independent of the strength of B_0. Suppression of this term, i.e., decoupling, is achieved by irradiating the sample continuously at the Larmor frequency of the proton spins. A measure for the required rf-amplitude, expressed in frequency units $\omega_1 = \gamma_H B_1$, is the strength of the carbon–proton or proton–proton dipole–dipole coupling, whichever is the stronger one. The critical aspect is the frequency match between the decoupling rf-field and the proton Larmor frequency.

While Thomas Klutz in our lab studied a powder sample of BEDT-TTF[1] [Bis(ethylen-dithio)tetrathiofulvalen] by MASS he made the observation that he

---

[1] BEDT-TTF forms electrically conducting radical salts with a number of counterions. Some of these salts become superconducting at $T \lesssim 10\,K$ [11]

had to hit the proton resonance within ± 2 ppm in order to be able to resolve in the MASS spectrum all of the carbon resonances of the $(CH_2)_2$ end groups of BEDT-TTF. The width of the proton resonance of BEDT-TTF is about 40 kHz or 150 ppm at $\nu_{oH} = 270$ MHz. Being surprised how sensitive the carbon NMR reacts on the frequency of the proton decoupling rf-field, he carried out a series of tests and did calculations on a model system. The model system was an isolated CH_2 group. An almost ideal realization of this model system is represented by the CH_2 group of a crystal of malonic acid, $^*CH_2(COOH)_2$ (*carbon enriched to 90%). The results of T. Klutz can be summarized as follows: *static sample*: crystal orientation such that in the Hamiltonian

$$\mathscr{H} = A_1 I_z^1 S_z + A_2 I_z^2 S_z + D[I_z^1 I_z^2 - 1/4(I_+^1 I_-^2 + I_-^1 I_+^2)] \\ + \omega_1(I_x^1 + I_x^2) + \Delta\omega(I_z^1 + I_z^2)$$

the coupling strengths were $A_1 = -2\pi.14$ kHz, $A_2 = 2\pi.20$ kHz, and $D = 0$, while the offset $\Delta\omega$ was zero. S_α denotes carbon, I_α^1, I_α^2 ($\alpha = x, y, z$) denote proton spin operator components. For $\omega_1 = 0$, a triplet is observed. The calculation for the model system reveals that the central component of the triplet is actually an unresolved doublet. When ω_1 is increased this doublet evolves into a triplet with the center component at the chemical shift of the carbon. The *intensity* of this center component increases with increasing ω_1. The intensities of all the outer components in the spectrum decrease with increasing ω_1. These components also move outwards in the spectrum. For $\omega_1 = 2\pi.21.8$ kHz, i.e., $\omega_1 \approx$ largest proton–carbon coupling strength $= A_2$ three quarters of the total intensity reside already in the central component, but it takes as much as $\omega_1 = 2\pi.50$ kHz $> 2A_2$ to shift 95% of the total intensity into this central component. We thus see that decoupling cannot be viewed as a reduction of coupling strengths. It has the effect of eventually shifting all the spectral intensity into one (the central component) of the spectrum.

The malonic acid crystal also allows for an orientation where $|D| \gg |A_1|, |A_2|$, i.e., where the proton–proton coupling is much larger than all the proton–carbon couplings. Klutz took a series of spectra with the crystal oriented such that $D = 2\pi.10$ kHz, $A_1 = 2\pi.0.73$ kHz, $A_2 = 2\pi.1.0$ kHz. For $\Delta\omega = 0$, $\omega_1 = 0$ an unstructured line is observed. It took $\omega_1 = |D|$, i.e., $\omega_1 \gg |A_1|, |A_2|$ to narrow the line down to the extent that its *amplitude* become 85% of the limiting value for very large values of ω_1.

Returning to the first crystal orientation ($D \approx 0$, $A_1 = -2\pi.14$ kHz, $A_2 = 2\pi.20$ kHz), Klutz studied also the offset dependence ($\Delta\omega \neq 0$) of the linewidth for $\omega_1 = 2\pi.32.6$ kHz, i.e., for a decoupling power level where on-resonance ($\Delta\omega = 0$) the line has reached about 90% of its limiting height. The interesting result obtained is that the linewidth increased already by roughly 50% when the offset $\Delta\omega$ was set to either $+5$ ppm or -5 ppm. When discussing the field dependence of decoupling the question arises whether a ppm scale, i.e., a scale proportional to B_o, or an *absolute* scale is appropriate. As the relevant quantities for comparison—the dipolar interaction strengths—are independent of B_o, an *absolute* scale seems to be the better choice. For the prospected field of 23.4 Tesla this implies that in a sample containing, say, aromatic as well as aliphatic carbons and protons, it will not be possible to decouple with straight cw irradiation the carbons from *all* protons simultaneously. This may sound as a disadvantage of very high field ^{13}C NMR and for routine applications it *is* a disadvantage but one should also not overlook the benefits of the situation: The linewidths in the carbon spectrum not only will quickly reveal the kind of protons to which the carbons are bound, but also give quick and accurate information about the chemical shifts of the protons.

Rotating sample: The above remarks about the static sample apply more-or-less to the rapidly rotating sample as well. We may even say that the problem with the offset dependence of the decoupling efficiency is more acute with spinning than with static samples. This is because the inherent linewidths are smaller for spinning samples, which means that a smaller residual dipolar broadening makes itself felt in the width or height of the resonance line. This is the starting observation of T. Klutz referred to above.

5.2 Spinning Rate ω_s

The quantity with which ω_s must be compared is the spread $\omega_{oc}\Delta\sigma$ of the carbon frequencies resulting from the chemical shift anisotropy $\Delta\sigma$. For $\omega_s \lesssim \omega_{oc}\Delta\sigma$ spinning sidebands are prominent, for $\omega_s \gg \omega_{oc}\Delta\sigma$ almost all of the spectral intensity resides in the center band at the isotropic shift of the carbon nucleus. For aliphatic, carboxylic, and aromatic protons the chemical shift anisotropies are, roughly, 30 ppm, 150 ppm, and 180 ppm [12]. At the presently preferred field of 7 Tesla, $\omega_{oc} = 2\pi.75$ MHz which implies that $\omega_{oc}\Delta\sigma$ becomes for aliphatic, carboxylic, and aromatic carbons $2\pi.2.25$ kHz, $2\pi.11.25$ kHz, and $2\pi.13.5$ kHz respectively. While spinning rates of $2\pi.10$ kHz are just becoming standard, many MASS experiments are successfully carried out with $\omega_s \lesssim 2\pi.5$ kHz. This means that spinning side bands are essentially avoided only for aliphatic carbons. For other carbons the sidebands must be suppressed with one of the available TOSS (total suppression of sidebands) techniques. This implies sacrificing signal intensity. On the other hand, good use can be made of the sidebands: the analysis of their intensities allows to determine the principal components of the chemical shift tensor which is valuable additional information.

Moving to still higher fields and frequencies—250 MHz for carbons, 1 GHz for protons—will lead to many additional sidebands even for aliphatic carbons, with all its drawbacks and benefits. Of course, the spinning rate can also be increased. To the best of my knowledge, spinning rates up to $2\pi.27$ kHz have already been demonstrated with rotor diameters of 4 mm. Sideband intensities at $\omega_{oc} = 2\pi.250$ MHz and $\omega_s = 2\pi.27$ kHz will be equal to those at $\omega_{oc} = 2\pi.75$ MHz and $\omega_s = 2\pi.8.1$ kHz and present experience shows that this latter is an acceptable spinning speed.

In essence, limited spinning rates and an increased number of spinning sidebands will not constitute a serious problem for very high field solid-state carbon NMR.

5.3 Cross Polarization

The standard procedure of proton–carbon cross-polarization is to spin-lock the protons and to establish the so-called Hahn-Hartmann condition: $\omega_{1\,\text{proton}} = \gamma_{\text{proton}} B_{1\,\text{proton}} = \gamma_{\text{carbon}} B_{1\,\text{carbon}} = \omega_{1\,\text{carbon}}$ on-resonance for both protons and carbons.

At "very high" fields the problem arises that due to chemical shift dispersion this condition cannot be satisfied simultaneously for all carbons in the spectrum. Let us consider a typical situation: Assume there are carbons in the sample with a chemical shift difference $\Delta\sigma$ of 180 ppm. For aromatic carbons this may result simply from chemical shift anisotropy. Assume further that $\omega_{1\,\text{proton}} = \omega_{1\,\text{carbon}} = 2\pi.50\,\text{kHz}$ which corresponds to a width of a $\pi/2$ pulse of $t_{\pi/2} = 5\,\mu s$—certainly a realistic value. If the frequency of the carbon transmitter is set on-resonance with one kind of carbons (e.g., at one edge of the powder spectrum), the other will be out of resonance by $\Delta\omega = \Delta\sigma\omega_{oc}$. The offset $\Delta\omega$ should be compared with $\omega_{1\,\text{carbon}}$. For ω_{oc} equal to $2\pi.75\,\text{MHz}$ and $2\pi.250\,\text{MHz}$ ("high" and "very high" field), $\Delta\omega$ will be equal to $2\pi.13.5\,\text{kHz}$ and $2\pi.45\,\text{kHz}$. The comparison of $\Delta\omega$ and $\omega_{1\,\text{carbon}}$ shows that for current "high" fields $\Delta\omega$ is still reasonably smaller than ω_1 which means that the entire powder spectrum of, say, an aromatic compound can be excited by cross-polarization essentially uniformly. At "very high" fields, on the other hand, $\Delta\omega \approx \omega_{1\,\text{carbon}}$, i.e., the on-resonance condition is severely violated at least for part of the carbons in the spectrum and the result is that the powder spectrum will be distorted due to non-uniform excitation.

This sounds like a disadvantage of very high field ^{13}C solid-state NMR but, again, I trust that NMR spectroscopists will find means to overcome this problem. I recall that M. Mehring in his monography [1] anyhow recommends far-off-resonance cross-polarization because of its higher efficiency despite the longer cross-polarization time constant.

Finally I would like to mention a problem of cross-polarization which arises in connection with MASS when very high spinning rates are used[2]. Very high spinning rates are desirable in ^{13}C NMR in very high fields as discussed above. Cross-polarization lives on the secular proton–carbon DD-interactions. Magic angle sample spinning tends to average out these interactions when the spinning rate becomes of the order of or larger than the proton–carbon-coupling strengths. These coupling strengths are typically of the order of a few tens of kHz (times 2π), see, for instance, the numbers quoted above for a $-CH_2$ group.

This means that spinning rates around 20 kHz *are* effective in suppressing the proton–carbon DD-interactions and are thus liable to render the cross-polarization mechanism ineffective.

All this means that we probably cannot work in a field of 23.4 Tesla without altering the standard experiments which work well in fields between 2 and 7 Tesla. However, the challenge to overcome problems posed by a new situation is just an essential element of the excitement of our profession and should not scare us.

---

[2] My attention was drawn to this problem by Dr. Förster from the Bruker company during a discussion on spinning rates made available by manufacturers of NMR equipment

6 Other Nuclei

Most of the "other nuclei" have small magnetogyric ratios and many of them possess a quadrupole moment. They are not studied routinely by solid-state NMR as are protons, deuterons, carbon-13, and, perhaps, silicon-29 and phosphorus-31 nuclei. Both the small magnetogyric ratios and the quadrupole interactions entail special consequences for very high field solid-state NMR. Here we shall consider two of these.

6.1 Probe Ringing

The availability of a very high magnetic field appears to be particularly attractive for the NMR of nuclei with small magnetogyric ratios because it brings their Larmor frequencies in a more familiar range. Higher frequencies are not only desirable from the point of view of NMR sensitivity but also make the dead-time problems notorious of solid-state NMR more readily tractable. Remember that the dead-time requirements are usually posed in absolute terms and do not scale with the strength of the field. At a higher frequency they can be satisfied with a higher quality factor of the probe circuit, which is a highly desirable circumstance.

There is, however, at least one physical reason why solid-state NMR at 46.5 MHz of, say, Ag-109 in a field of 23.4 Tesla is much more tricky than is proton NMR at the same frequency but in the correspondingly lower field. This reason is (acoustic?) ringing of the probe which is excited by the rf-pulses.

The pertinent observations of many experimentalists who have struggled both in research and manufacturers' laboratories with low-frequency solid-state NMR are as follows: The response to a rf-pulse of the probe at ambient temperature and outside of the magnet is as expected, i.e., essentially exponential all the way down to the thermal noise level. The time constant of the decay of the pulse is, depending on the electrical Q of the circuit, in the range of a fraction to a few microseconds. Once the probe is inserted into the magnet, this no longer is so. The pulse is now followed by a slowly decaying disturbance which is referred to as acoustic ringing. While the strength and appearance of the ringing does depend on the frequency of the pulse, it is usually not at all monochromatic. Nevertheless it is coherent with the rf-carrier. This means that it grows up during accumulation, although after a single shot it may be below the thermal noise level. It also means that it cannot be circumvented by phase cycling. For a given frequency, the ringing is in general the stronger, the stronger the magnetic field is. It also increases when the temperature of the probe is lowered to the region of 100 K or lower. Notice that in solid-state NMR such low temperatures are often of particular interest.

The exact origin of the disturbances in question is only poorly understood. The crucial role of the magnetic field strongly suggests that the Lorentz force is involved. It exerts an alternating force on all conductors through which the rf-current flows, in particular on the wire of the NMR coil. This force may excite mechanical (acoustical) vibrations, hence the term acoustic ringing. The frequency of the driving force (= Larmor frequency) is usually much higher than the eigenfrequencies of all conceivable acoustic modes of the various structural elements of the probe. Let us take an example: The sound velocity in metals like copper is roughly 5000 m/s. The wavelength λ for, e.g., $\nu = 25$ MHz, is then 0.2 mm which is probably less than even the diameter of the wire used to wind the coil. Nevertheless a multitude of overtones may still be excited. The decay time of these mechanical oscillations is of the order of tens or even hundreds of microseconds and thus much longer than the decay of any purely electrical mode of the probe. The presence of the magnetic field provides, of course, also for a coupling of the mechanical vibrations back into the electrical circuit. The observed temperature dependence of the ringing is consistent with this description of its origin: The quality factor of mechanical vibrations tends to rise with decreasing temperature and a higher quality factor favors larger, longer-lasting excitations by rf-pulses, i.e., larger longer-lasting ringing.

Another source of the annoying disturbances may be the alternating eddy currents in the rf- shield of the probe or simply in the walls of the cryostat. The Lorentz force again provides for a coupling to and from mechanical modes. The increasing conductivity of the shields and walls at lower temperatures will increase the effect, as is observed.

People have tried various strategies against these hideous disturbances. One of them is to cut narrow slots into the rf-shields. The idea is to interrupt the paths of the suspected eddy currents. But neither this nor other suggested measures are a sure solution to the problem. Sometimes the best thing to do is to live with the ringing and to work with echoes. The pulse timing must then be chosen such that the ringing excited by the last pulse of the echo sequence has essentially decayed prior to the event of the echo maximum.

In any event the problem of acoustic ringing becomes more acute the lower the frequency and the higher the magnetic field is. I anticipate, therefore, that solid-state NMR of low-magnetogyric nuclei in very high fields will not be so straightforward as it may look at first sight but will require a great deal of difficult, zealous, and patient experimentation.

6.2 NMR of Quadrupolar Half-Integer Spins in High Fields

Most of the stable quadrupolar isotopes have half-integer spins. The few notable exceptions are ^{2}H, ^{6}Li, ^{10}B, ^{14}N, and ^{50}V. If the nuclear quadrupole-EFG interaction can be treated as a (small) pertubation of the Zeeman interaction—a likely situation in very high fields—the single-crystal NMR spectra of half-integer nuclei consist of the so-called central $\Delta m = \pm 1$ transitions between the $m = +1/2$ and $m = -1/2$ spin states, and satellites, i.e., transitions between $m = +3/2$ and $m = +1/2$ states, etc. The displacements of the satellites from the pure Larmor frequency ν_L are proportional to the quadrupole interaction, whereas the central transition is only shifted in second order. The satellites are often hard to observe and more often than not they are not observable at all because even slight crystal imperfections result in a distribution of EFGs and hence in an inhomogeneous broadening of the satellites. This broadening smears out the spectral intensity of the satellites over a broad frequency range with an accompanying loss of signal-to-noise. Measurements of EFG (to be precise: of quadrupole coupling-) tensors must then rely on the second-order shifted central transitions whose sharpness is much more tolerant against crystal imperfections. Now, the second-order shift of the central line:

$$\delta\nu^{(2)} = -\frac{\pi \cdot \nu_Q^2}{8\gamma B_0}[I(I+1) - 3/4](1 - \cos^2\Theta)(9\cos^2\Theta - 1)$$

scales with B_0^{-1}. In this expression axial symmetry of the EFG is assumed and:

$$\nu_Q = \frac{3e^2qQ}{2I(I-1)h} \quad \text{and} \quad \Theta = \sphericalangle(\vec{B}, \text{symmetry axis of EFG})$$

This means that in measurements of quadrupole coupling tensors via the second-order shifts of the central transition the size of the measurable effect ($\delta\nu^{(2)}$) decreases relative to the linewidth when the field is increased. Notice that this is just the opposite situation as in measurements of chemical shifts. For large quadrupole interactions, the decrease of $\delta\nu^{(2)}$ with increasing field strength may be a desirable effect, however, the opposite situation may arise as well. To give an example: consider the ^{39}K resonances in $KLiSO_4$ studied by B. Topic in our lab in a field of 8.36 T where ν_L of ^{39}K is 16.36 MHz [13]. ^{39}K is a spin 3/2 nucleus. The crystal quality was such that the satellites could not be observed. $KLiSO_4$ undergoes two successive phase transitions at T = 217 K and T = 190 K. The room-temperature phase is hexagonal and only one resonance is observed for $\vec{B}_0$ perpendicular to the hexagonal axis. The change of the ^{39}K spectrum at the first phase transition (T = 217 K) is of the order of the width of the line, but clearly detectable for B_0 = 8.36 T (cf. Fig. 4 of Ref. [13]). Now suppose the experiment had been done at B_0 = 23.4 T, in principle an attractive proposition. However, the second-order shifts of the central transitions of the structurally discernible potassium nuclei in the intermediate phase would now have been scaled down by a factor of three, i.e., they would have become so small that almost certainly they would have escaped detection.

This demonstrates, once again, that the choice of the strength of the field in solid-state NMR must be carefully adjusted to the physical problem at hand. The dream of the solid-state NMR spectroscopist who has not specialized on one particular nucleus and one particular class of materials is really the access to the full range of fields starting from actually zero to the very highest possible one.

7 References

1. Mehring M (1983) in: Diehl P, Fluck E, Kosfeld R (eds) NMR Basic principles and progress, vol 11, Springer-Verlag, Berlin, Heidelberg, New York
2. Fyfe CA (1983) Solid state NMR for chemists, CFC Press Guelph Ontario Canada
3. Gerstein B, Dybowski C (1985) Transient techniques in NMR of solids, Academic Press, New York
4. Blümich B, Spieß HW (1988) Angew. Chem. Int. Ed. Engl. 27: 1655
5. Sagnowski SF, Aravamudhan S, Haeberlen U (1977) J. Magn. Reson. 28: 271

6. Meier BH (1984) Dynamik und Struktur der intermolekularen Wasserstoffbrücken in dimeren Carbonsäuren untersucht mit Festkörperkernresonanz und Neutronenbeugung. Thesis, ETH Zürich
7. LeFloch-Durand M, Haeberlen U, Müller C (1986) J. de Physique 43: 107
8. Benz S, Haeberlen U, Tegenfeldt J (1986) J. Magn. Reson. 66: 125
9. Slosarek G, Heuer A, Zimmermann H, Haeberlen U (1989) J. Phys.: Condens. Matter 1: 5931
10. Müller C, Schajor W, Zimmermann H, Haeberlen U (1984) J. Magn. Reson. 56: 235
11. Gärtner S, Gogu E, Heinen I, Keller HJ, Klutz T, Schweitzer D (1988) Solid State Comm. 65: 1531
12. Spieß HW (1978) In: Diehl P, Fluck E, Kosfeld R (eds) NMR Basic principles and progress, vol 15, Springer-Verlag, Berlin, Heidelberg, New York
13. Topic B, Haeberlen U, Blinc R (1988) Z. Phys. B Condensed Matter 70: 95

Author Index Volumes 21–25